RESILIENT LANDSCAPE
VISION
FOR LOWER WALNUT CREEK

Baseline Information & Management Strategies

NOVEMBER 2016

SFEI | **AQUATIC SCIENCE CENTER**
SAN FRANCISCO ESTUARY INSTITUTE & THE AQUATIC SCIENCE CENTER

PRIMARY AUTHORS
Scott Dusterhoff
Carolyn Doehring
Sean Baumgarten
Robin Grossinger

DESIGN AND LAYOUT
Ruth Askevold

PREPARED BY   San Francisco Estuary Institute-Aquatic Science Center

IN COOPERATION WITH
San Francisco Estuary Partnership
San Francisco Bay Conservation and Development Commission
San Francisco Bay Joint Venture

FUNDED BY   San Francisco Bay Water Quality Improvement Fund, EPA Region IX
Contra Costa County Flood Control & Water Conservation District

A PRODUCT OF FLOOD CONTROL 2.0

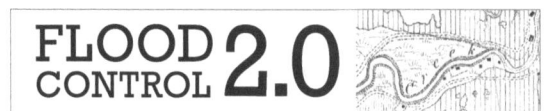

FLOOD CONTROL **2.0**

SAN FRANCISCO ESTUARY INSTITUTE PUBLICATION #782

SUGGESTED CITATION

San Francisco Estuary Institute-Aquatic Science Center. 2016. *Resilient Landscape Vision for Lower Walnut Creek: Baseline Information & Management Strategies.* A SFEI-ASC Resilient Landscape Program report developed in cooperation with the Flood Control 2.0 Regional Science Advisors and Contra Costa County Flood Control and Water Conservation District, Publication #782, San Francisco Estuary Institute-Aquatic Science Center, Richmond, CA.

Version 1.0 (November 2016)

REPORT AVAILABILITY

Report is available at floodcontrol.sfei.org

COVER CREDITS

Front cover, left to right: map of historical conditions of lower Walnut Creek, developed for the Flood Control 2.0 project; imagery courtesy of Google Earth.

Back cover, top to bottom: map of possible measure to sustain resilient marshes, developed for the Flood Control 2.0 project; map of modern conditions of lower Walnut Creek, developed for the Flood Control 2.0 project; 1895 lithograph of Diablo Valley shows Walnut Creek through the marsh (courtesy of Dean McLeod and Contra Costa County Historical Society); photograph of Walnut Creek by Carolyn Doehring (SFEI).

CONTENTS

Executive Summary   2

Introduction   4

Flood Control 2.0   6

Recent Management of Walnut Creek   8

The Process for Developing a Long-Term Vision for Lower Walnut Creek   10

Historical Ecology of Lower Walnut Creek   12

The Lower Walnut Creek Landscape (maps, circa 1850 and circa 2010)   16

Landscape Change of Lower Walnut Creek   18

Sediment Accumulation in Lower Walnut Creek   20

Changing Future Conditions   24

Vision Guidance for Improving Landscape Functionality and Resilience   26

Lower Walnut Creek Vision Strategies   29

Lower Walnut Creek Vision Measures   40

Constraints   42

Timing of Implementation   43

Moving Forward   44

Acknowledgements   45

References   46

EXECUTIVE SUMMARY

Lower Walnut Creek (Contra Costa County, CA) and its surrounding landscape have undergone considerable land reclamation and development since the mid-nineteenth century. In 1965, the lower 22 miles of Walnut Creek and the lower reaches of major tributaries were converted to flood control channels to protect the surrounding developed land. In the recent past, sediment was periodically removed from the lower Walnut Creek Flood Control Channel to provide flow capacity and necessary flood protection. Due to the wildlife impacts and costs associated with this practice, the Contra Costa County Flood Control and Water Conservation District (District) is now seeking a new channel management approach that works with natural processes and benefits people and wildlife in a cost-effective manner. Flood Control 2.0 project scientists and a Regional Science Advisory Team (RSAT) worked with the District to develop a long-term management Vision for lower Walnut Creek that could result in a multi-benefit landscape that restores lost habitat and is resilient under a changing climate.

Development of the Vision began with technical analyses focused on understanding past and present landscape functioning and the changes to key processes and landscape features over the past 150 to 200 years. The key findings from these analyses include the following:

- **Landscape change** – During the mid-19th century, lower Walnut Creek was surrounded by a continuous expanse of tidal wetlands occupying approximately 5,000 acres. Perennial freshwater marsh, willow thicket, and alkali meadow formed a large non-tidal wetland complex that adjoined the tidal marsh on its southern end. Over the past century and a half, the non-tidal wetland complex has been lost completely and the tidal wetland area has decreased by 40%. The remaining marsh areas are highly fragmented and cut-off from Walnut Creek by engineered levees. The loss of tidal wetland area has resulted in a substantial decrease in tidal prism and contributed to current in-channel sediment accumulation issues. Around the mouth of Walnut Creek, sediment accumulation has caused the position of the shoreline to expand into the Bay by up to half a mile.

- **Sediment accumulation** – Since 1965, approximately 1.4M cubic yards of sediment have been removed from lower Walnut Creek, with approximately 70% coming from the tidal zone downstream of head of tide (i.e., the inland extent of tidal inundation at mean higher high water). Repeat channel cross-section surveys indicate that the channel quickly fills in with sediment following dredging events (due in large part to decreased tidal prism and associated channel scour capacity) and that the channel bed elevation is currently at or near "quasi-equilibrium." A channel sediment budget for 1965-2007 indicates that approximately 80% of the watershed-derived sediment made it through lower Walnut Creek and out to the Bay, while the remaining 20% was likely deposited directly following channel construction and subsequent dredging events.

In November 2015, Flood Control 2.0 scientists, in partnership with the District, held a workshop to present the findings from the technical analyses and develop multi-benefit management concepts that form a long-term Vision for lower Walnut Creek. The workshop participants included the RSAT and other local organizations involved in flood risk management, baylands management, permitting, and water quality. Based on the information presented at the workshop and an extensive knowledge of regional habitat needs, the RSAT recommended the following suite of management strategies and associated actions (or measures) throughout lower Walnut Creek:

- **Strategy 1 – Sustaining resilient marshes by improving natural delivery of freshwater and sediment**

  Measure 1 - Set back levees

  Measure 2 - Reconnect creeks to floodplains

  Measure 3 - Create zones for distributary corridors

  Measure 4 - Modify transportation and pipeline infrastructure

- **Strategy 2 – Sustaining resilient marshes using dredged sediment**

  Measure 5 - Maintain marsh elevations with dredged sediment

  Measure 6 - Protect marsh edge with dredged sediment

- **Strategy 3 – Sustaining resilient marshes using treated wastewater**

  Measure 7 - Support freshwater wetlands with wastewater discharges

  Measure 8 - Support seepage slopes with diffuse wastewater discharges

- **Strategy 4 – Improving ecological connectivity across marshes and along creeks**

  Measure 9 - Enhance wildlife corridors

  Measure 10 - Protect and restore transition zones

The next steps for implementing these measures include conducting feasibility analyses, collaboration with local landowners, garnering regulatory agency support, and securing the necessary funds. The pace of sea-level rise combined with the overall lead time needed to implement the management measures suggest that planning should begin in the near future so that implementation can occur before sea level has risen so high that it's too late.

Pacheco Marsh. (photograph courtesy of Stephen Joseph Photography, stephenjosephphoto.com)

INTRODUCTION

Flood control channels in the Bay Area are the subject of increasing concerns about aging infrastructure, regulatory restrictions, ongoing maintenance issues, and the challenge of increasing water levels with sea-level rise. In addition, there is a growing need to use sediment trapped in the channels as a resource to build and maintain tidal marsh elevations. Here, we present a possible future vision for lower Walnut Creek and adjacent baylands that includes several components that would restore and support natural processes, and in turn, benefit aspects of flood risk management and ecosystem functioning. The Lower Walnut Creek Vision is an element of an EPA-funded project called Flood Control 2.0, which is aimed at integrating wildlife habitat improvement and flood risk management along the San Francisco Bay shoreline for the 21st century and beyond.

Over the past two centuries, lower Walnut Creek and its surrounding landscape have undergone considerable land reclamation and development, which has led to significant channel modifications and a dramatic loss of wildlife habitat. With ongoing sedimentation of the channel and accelerating sea-level rise, channel conveyance for flood flows is a rising concern. In the past, the channel was dredged to increase flow capacity; however, the Contra Costa County Flood Control and Water Conservation District (District) is currently re-evaluating its management practices, especially given dredging impacts to wildlife and future challenges with climate change. Through its current Lower Walnut Creek Restoration Project (LWC Restoration Project), the District is seeking to build and manage a sustainable channel that provides critical flood protection in a way that is more compatible with the plants and wildlife that call the Creek home. Flood Control 2.0 team members and project partners worked with the District to explore a range of landscape-scale opportunities for integrating ecological benefits with flood risk management on lower Walnut Creek considering likely impacts of climate change.

Pacheco Marsh.
(photograph courtesy of Stephen Joseph Photography,
stephenjosephphoto.com)

The process for developing the Vision had three main elements. First, the San Francisco Estuary Institute (SFEI) built a baseline understanding of the historical and contemporary geomorphic and ecological conditions, and assessed the likely impact of future drivers (particularly sea-level rise). Second, these findings were presented at a workshop held in November 2015 by the Flood Control 2.0 project team with the District and a Regional Science Advisory Team (RSAT) made up of regional experts. The goal of the workshop was to explore potential integration of improved ecosystem health and flood risk management on lower Walnut Creek to envision a landscape with increased resilience of ecosystem services and ecological functions to climate change. The visioning workshop considered Walnut Creek and the adjacent floodplains from Concord Avenue downstream to Suisun Bay, including land beyond the District's jurisdiction (hence forth referred to as the "study area"). Third, the ideas presented at the workshop and developed in follow-up discussions with the RSAT were synthesized into four over-arching strategies: 1) Sustaining Resilient Marshes by Improving Natural Delivery of Freshwater and Sediment; 2) Sustaining Resilient Marshes Using Dredged Sediment; 3) Sustaining Resilient Marshes Using Treated Wastewater; and 4) Improving Ecological Connectivity Across Marshes and Along Creeks.

The Lower Walnut Creek Vision is intended to help the District, partner agencies, landowners, and other stakeholders explore approaches for multi-benefit landscape management in the coming decades. Lower Walnut Creek has the potential to be redesigned as a better-functioning estuary-delta system that is more resilient to future climate change impacts (e.g., a sea-level rise, salinity shifts) while providing the desired level of flood risk management and improved habitat conditions. Ideally, the Vision will continue to be refined through subsequent analyses and coordinated with visions for landscape management in the watersheds that drain to lower Walnut Creek. This Vision can also be used to help guide channel-bayland redesign efforts around the Bay in similar landscapes.

FLOOD CONTROL 2.0

Over the past 200 years, many of the creeks that drain to San Francisco Bay have been modified for flood risk management. Channel modifications include the building of concrete trapezoidal channels, construction of levees along channels, and complete realignment. In many instances, these flood risk management actions have had considerable impacts on geomorphic channel processes and ecological functioning and the way that sediment and water pass from the watershed to the Bay. Historically, creeks frequently transported watershed-derived sediment to the baylands. Now, leveed channels (with reduced tidal prism) trap sediment at the Bay interface. This has resulted in excess channel deposition, frequent channel dredging, and subsequent adverse impacts to resident plants and animals. Local agencies that operate and maintain flood control channels are coming under increasing pressure from resource agencies to manage or redesign flood infrastructure to provide beneficial uses beyond flood conveyance, including habitat for rare, threatened, or endangered species. In addition, sediment trapped in flood control channels is now being seen as a valuable commodity for baylands restoration.

Recognizing the environmental impacts associated with current flood risk management activities, the high cost of maintaining aging infrastructure, the challenges associated with maintaining flood conveyance in the face of a rising sea level, and the high value of dredged sediment, flood control managers and regulatory agencies are calling for a new overall approach for channel management.

Flood Control 2.0 is an innovative regional project that seeks to integrate habitat improvement and flood risk management at the Bay interface. The project focuses on helping flood control agencies and their partners create landscape designs that promote improved sediment transport through flood control channels, improved flood conveyance, and the restoration and creation of resilient bayland habitats. In addition, the project focuses on beneficial re-use options for dredged sediment from highly constrained flood control channels with limited restoration opportunities. Through a series of coordinated technical, economic, and regulatory analyses, Flood Control 2.0 addresses some of the major elements associated with multi-benefit channel design and management at the Bay interface and will provide critical information that can be used to develop long-term solutions that benefit people and habitats.

Findings from this report and other creek studies (e.g., San Francisquito Creek, Novato Creek) will be synthesized into an online "toolbox." The toolbox will include channel classifications and relevant management concepts (e.g., creek-bayland connections, beach nourishment), a "marketplace" for baylands restoration practitioners to find available dredged sediment (Sedi-Match), a regulatory guidance document with case studies for the regulatory issues associated with flood control project elements (e.g., impacts to existing wetlands, sediment re-use), and a benefit-cost analysis. The toolbox will be completed and available to the public in 2016. In combination with other regional plans (e.g., Baylands Ecosystem Habitat Goals Science Update), this project will provide information to flood control managers and the restoration community for planning sustainable, long-term, multi-benefit redesign projects given landscape, regulatory, and economic challenges.

Additional project information: floodcontrol.sfei.org

PROJECT LEADS

- SAN FRANCISCO ESTUARY PARTNERSHIP (SFEP)
- SAN FRANCISCO ESTUARY INSTITUTE (SFEI)
- SAN FRANCISCO BAY CONSERVATION AND DEVELOPMENT COMMISSION (BCDC)
- SAN FRANCISCO BAY JOINT VENTURE (SFBJV)

PROJECT PARTNERS

- CONTRA COSTA COUNTY FLOOD CONTROL AND WATER CONSERVATION DISTRICT (CCCFCWCD)
- MARIN COUNTY DEPARTMENT OF PUBLIC WORKS (MCDPW)
- SAN FRANCISQUITO CREEK JOINT POWERS AUTHORITY (SFCJPA)

REGIONAL SCIENCE ADVISORY TEAM

- PETER BAYE (ECOLOGICAL CONSULTANT)
- LETITIA GRENIER (SFEI)
- JEFF HALTINER (HYDROLOGY CONSULTANT)
- JEREMY LOWE (SFEI)

FLOOD CONTROL 2.0 PROJECT STRUCTURE

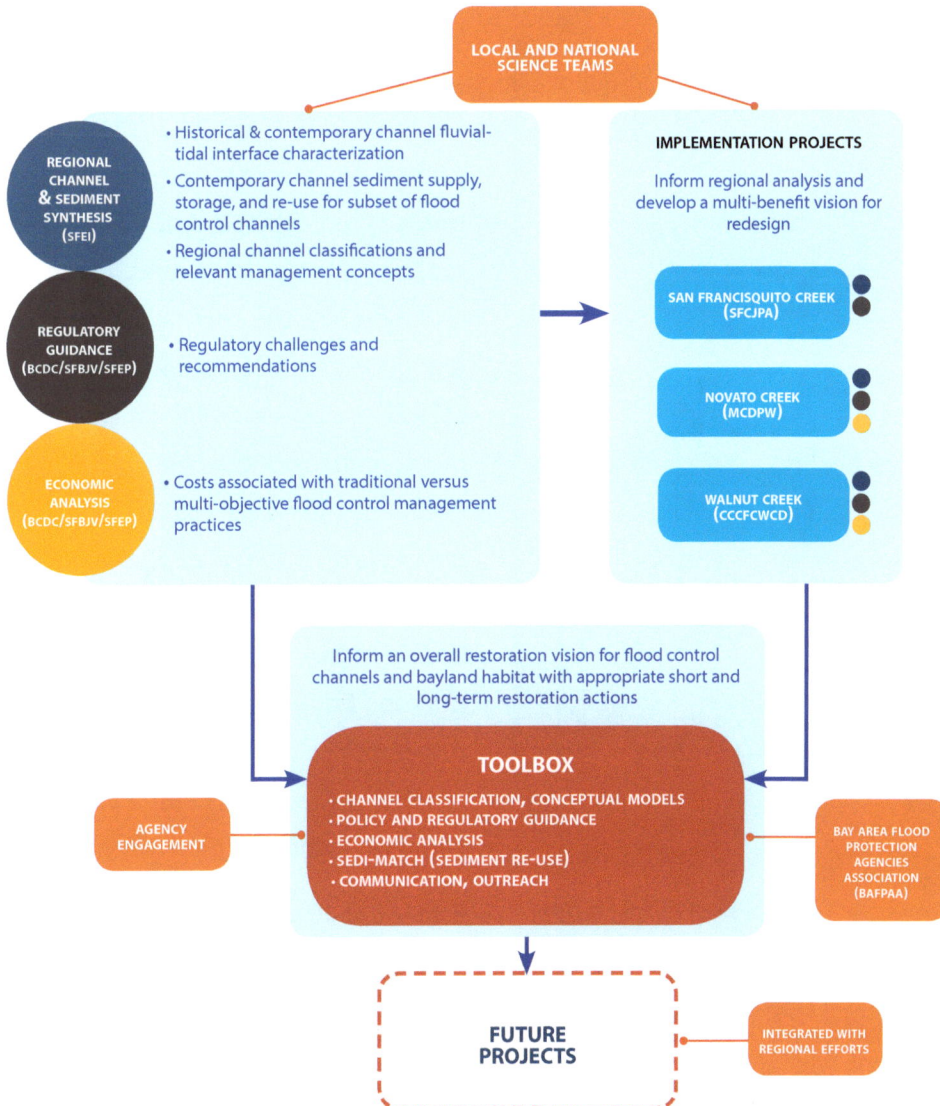

RECENT MANAGEMENT
OF WALNUT CREEK

Walnut Creek in central Contra Costa County flows through the cities of Walnut Creek, Pleasant Hill, and Concord before entering Suisun Bay. Over the past 150 years, the watershed has undergone many changes. Historically, Walnut Creek would spread out onto a large freshwater marsh before draining to a tidal marsh. During the late 1800s and early 1900s, the watershed experienced several damaging floods with high sediment loads, leading to levee building and channel realignment in the creek's lower reaches. In the 1960s, the United States Army Corps of Engineers (USACE) built the Walnut Creek project, which included the lower 22-mile reach of mainstem Walnut Creek and the lower reaches of major tributaries. In the tidal portion of the project, the channels were widened and flood control levees were constructed. USACE anticipated minimal long-term channel maintenance but the channel filled with sediment soon after construction. Less than a decade after it was built, USACE dredged over 800,000 cubic yards of newly deposited sediment from the mouth to the BNSF Railroad bridge before transferring channel maintenance to the District. Sedimentation of Walnut Creek has continued, reducing the channel's capacity to convey floodwaters.

For the last 40 years, the District has maintained the Walnut Creek project in accordance with USACE standards with the exception of dredging the tidal portion of the project. Dredging to meet USACE

Lower Walnut Creek study area. (Imagery courtesy of NAIP 2014)

8

requirements would have negative impacts on aquatic and wetland habitat and therefore be extremely difficult to permit and costly to mitigate. The District determined that dredging was not sustainable and worked with USACE from 2004 to 2012 to find a sustainable solution; however, the effort stalled due to lack of federal funding. In 2014, Congress approved the District's request for a selective deauthorization from USACE project authority that returned oversight of 2.5 miles of lower Walnut Creek and 1.5 miles of lower Pacheco Creek to the District.

The District is currently undergoing an extensive community-based planning process to develop restoration alternatives within the District's existing jurisdiction (called the Lower Walnut Creek Restoration Project). The Restoration Project seeks to transform the channel into a sustainable system that provides wildlife habitat and flood protection benefits, and has reasonable maintenance costs (CCCFCWCD 2014). This effort has roots in the District's "50 Year Plan," which is a vision for converting classic engineered flood control channels back to natural channels that provide the same level of flood protection throughout Contra Costa County by mid-century (CCCFCWCD 2009). This effort also includes close coordination with the restoration plan being developed by the District and the John Muir Land Trust for Pacheco Marsh (122 acres) adjacent to the Restoration Project at the mouth of Walnut Creek.

This history has created an array of challenges and opportunities associated with channel management:

CHALLENGES

- Transportation and transmission infrastructure (e.g., Union Pacific and BNSF railroads, Waterfront Road, pipelines) within the floodplain

- The need to maintain access to infrastructure (e.g., Central Contra Costa Sanitary District [CCCSD] outfall) for maintenance and repairs

- Landfills and contaminated sites

- Limited lands that are under District ownership; many adjacent lands are privately owned

- High maintenance costs and regulatory restrictions for continued dredging

- Limited areas for habitat migration with sea-level rise, because much of the floodplain is developed or is naturally at higher elevations

- Risk of tidal marsh loss (likely changing to mudflat or subtidal habitat) with sea-level rise

- Potential that watershed sediment supply may not be able to support baylands under rising sea levels

OPPORTUNITIES

- Improve level of flood risk protection through setback of levees, floodplain expansion, and re-connection to fluvial channels

- Increase tidal prism to maintain flood conveyance and promote sedimentation on the marsh plain

- Re-use sediment to enhance resilience of tidal habitats and habitat value of estuarine-terrestrial transition zones

- Minimize or avoid environmental disturbances and costs associated with maintenance channel dredging

- Utilize treated wastewater for enhancing salinity gradients across marsh habitats

- Restore and enhance habitat for wildlife (e.g., Ridgway's Rail, salt marsh harvest mouse), including tidal marshes, fluvial floodplains, transition zones, and riparian areas

- Increase public access for recreation and wildlife viewing

THE PROCESS
FOR DEVELOPING A LONG-TERM VISION FOR LOWER WALNUT CREEK

Step 1
Pre-Workshop

UNDERSTANDING LANDSCAPE FUNCTIONING

Developing management approaches that lead to a resilient landscape requires understanding the processes that create and maintain landforms and associated habitat types. The first step in developing this understanding included synthesizing archival data to reconstruct the pre-development (mid-19th century) landscape of lower Walnut Creek and adjacent baylands. The historical conditions were then compared to contemporary conditions to highlight changes in physical processes and habitat extent and configuration. SFEI also drew on existing data to complete a contemporary geomorphic analysis of lower Walnut Creek focusing on the magnitude of watershed sediment yield and the major drivers for the current excess sedimentation issues.

Step 2
At Workshop

WORKSHOP

In November 2015, Flood Control 2.0 project leads, in partnership with the District, held a workshop to discuss ideas for improving flood management and habitat conditions within lower Walnut Creek. A Regional Science Advisory Team (RSAT) consisting of regional experts with backgrounds in flood risk management, tidal marsh ecology, and coastal geomorphology were recruited to review the current challenges facing the District and identify potential strategies to address these challenges. Local organizations involved with flood risk management, baylands management, permitting, and water quality were also present at the workshop. The workshop was facilitated by Andy Gunther of the Bay Area Ecosystems Climate Change Consortium (BAECCC).

During the workshop, SFEI presented an analysis of historical and contemporary channel morphology and alignment, sediment dynamics, and habitat extent and configuration in the study area. The District and Environmental Science Associates (ESA; engineering consultants

working on the lower Walnut Creek Restoration Project) presented the history of flood management in lower Walnut Creek, existing infrastructure, and the District's current efforts for enhancing flood management and habitat benefits through near-term restoration projects. During the workshop field trip, participants visited Pacheco Marsh and the lower reaches of Walnut Creek and Pacheco Creek. Based on the information presented, the RSAT provided their expert advice on a range of multi-benefit opportunities for landscape change.

Step 3
Post-Workshop

DEVELOPING THE VISION

The ideas developed at the workshop were synthesized into four overarching strategies aimed at improving long-term resilience of the lower Walnut Creek landscape to support ecosystem services and wildlife habitat under changing future conditions. Each strategy contains several detailed measures (i.e., management actions) that focus on both physical and ecological enhancements to the study area. Together, the measures make up a long-term landscape "Vision."

We used SFEI's recently released Landscape Resilience Framework (Beller et al. 2015) to help guide Vision development. Within this framework, landscape resilience is defined as "the ability of a landscape to sustain desired ecological functions, robust native biodiversity, and critical landscape processes over time, under changing conditions, and despite multiple stressors and uncertainties." The framework provides a robust guide for incorporating the fundamental drivers of ecological resilience into the design of ecosystems and environmental management at the landscape scale. Additional information about the Landscape Resilience Framework can be found at resilientsv.sfei.org.

Walnut Creek Vision Workshop. (photographs by SFEI, 2015)

HISTORICAL ECOLOGY
OF LOWER WALNUT CREEK

OVERVIEW

Designing a resilient landscape requires reestablishing the processes that allow ecosystems to thrive, recover, and adapt under changing conditions while providing benefits like flood protection and erosion control. One of the most useful ways to identify those processes and learn how they can be reestablished is to study how a given landscape looked and functioned prior to its extensive modification: its historical ecology.

The use of historical data to study past ecosystem characteristics is a powerful tool not only for reconstructing the past landscape, but also for revealing patterns and processes still operating today and for helping us to envision future landscape potential. Reconstructing the historical ecology of lower Walnut Creek can shed light on a range of important questions: What was the distribution and extent of wetland habitat types? What wildlife species relied on these habitats? How did water and sediment move across the landscape? How has the landscape been modified over the past 150 years? What physical processes or remnant features are still intact that might provide opportunities for restoring ecological functions and enhancing landscape resilience?

To address these and other questions, we collected and synthesized a range of historical sources to reconstruct how the lower Walnut Creek landscape looked and functioned in the recent past. The study area for the historical reconstruction (page 16) encompasses the full historical extent of the tidal wetlands area (i.e., the "baylands") around lower Walnut Creek, adjacent non-tidal wetlands, and downstream portions of the major stream channels that flowed into the baylands; it extends along the shoreline from Bulls Head Point (at the southern landing of the Benicia Bridge) east to Seal Bluff Landing (near Port Chicago).

Data Collection and Compilation

We drew on a wide variety of archival datasets dating back to the late 18th century, including maps, photographs, drawings, and textual documents. Key sources included Spanish explorer diaries, U.S. Coast [and Geodetic] Survey maps, General Land Office Survey plats and field notes, newspaper articles, USGS topographic quadrangles, USDA soil surveys, and aerial photographs. Data were collected from online databases and local, regional, state, and federal archives (see table below).

Source Institution	Location
The Bancroft Library, UC Berkeley	Berkeley
Bureau of Land Management (remotely)	Sacramento
California State Archives	Sacramento
California Historical Society	San Francisco
California State Railroad Museum	Sacramento
Contra Costa County Historical Society	Martinez
Earth Sciences and Map Library, UC Berkeley	Berkeley
Society of California Pioneers	San Francisco
Water Resources Collections and Archives, UC Riverside	Riverside

Source institutions visited or contacted.

Selected data sources were compiled in a Geographic Information Systems (GIS) database. We georeferenced approximately 40 maps and over 100 spatially explicit excerpts from textual documents. A photomosaic of orthorectified 1939 aerial imagery (USDA 1939), covering all of Contra Costa County, was published by SFEI in 2011 (Salomon 2011); orthorectification of the existing photomosaic was refined in some portions of the study area to improve local accuracy. We also orthorectified two aerial photographs from 1928-9 (Russell 1928-9).

Synthesis and Mapping

Historical sources differ widely in terms of accuracy, level of detail, spatial extent, and purpose. While no single source provides a complete picture of the historical landscape, the comparison and synthesis of multiple independent sources allows for a much more accurate reconstruction. Data sources assembled for this study were synthesized to develop a series of GIS layers representing average ecological conditions circa 1850, prior to major Euro-American modifications (page 16). Features were classified as one of the following habitat types:

Tidal Marsh — vegetated portions of the baylands

Marsh Pond/Panne — open water or unvegetated areas on the marsh plain

Subtidal Channel — portions of tidal channels that do not completely drain at low tide

Channel Flat and Bay Flat — portions of tidal channels and the bay that dewater during low tide. Small tidal sloughs were mapped as line features

Perennial Freshwater Wetland / Willow Thicket — non-tidal wetlands dominated by emergent vegetation and willows

Alkali Meadow — seasonal wetlands characterized by moderately alkaline soils, seasonal flooding, and a salt-tolerant plant community

Stream Channels/Distributaries — non-tidal stream channels. Streams were mapped as line features, and streams that spread into multiple distributary channels are shown with a forked crow's foot symbol

Feature boundaries were mapped from the most spatially accurate sources representative of pre-modification conditions. Wherever possible, the classification and extent of each feature was verified using secondary sources. Each feature was attributed in GIS with both supporting sources and certainty levels representing our confidence in feature classification, shape/size, and location.

For the landscape change analysis (pages 18-19), modern habitat type mapping (page 17) was compiled from the Bay Area Aquatic Resource Inventory baylands and wetlands layers (SFEI-ASC 2015), a Contra Costa County stream layer (Contra Costa County 2008), and the San Francisco Bay Shore Inventory dataset (SFEI 2016).

Maps, photographs, and textual documents comprised the principal data types collected. (left to right, top to bottom: Rodgers 1856, courtesy of NOAA; Hesse 1861, courtesy of The Bancroft Library, UC Berkeley; BANC MSS Land Case Files 87 ND, courtesy of The Bancroft Library, UC Berkeley; Russell 1928-9, courtesy of Earth Sciences & Map Library, UC Berkeley)

HISTORICAL ECOLOGY
OF LOWER WALNUT CREEK
THE LOWER WALNUT CREEK LANDSCAPE, CIRCA 1850

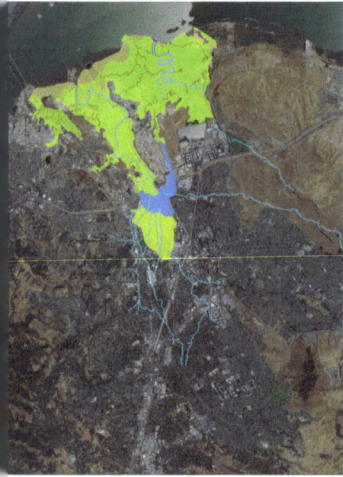

During the mid-19th century, the lower Walnut Creek watershed was dominated by extensive wetlands, meandering creeks, and grassy plains. The following section describes key features of the historical landscape.

Tidal Marsh

Low-lying areas along lower Walnut Creek supported a large tidal marsh extending from Suisun Bay south to present-day Highway 4 (Rodgers 1856, Rodgers and Chase 1866). The marsh is described as a "tule marsh" in some accounts (e.g., Ransom 1851, Coffee 1857), and as a "salt marsh" in others (e.g., Taylor 1864a); early surveyors also noted the presence of "samphire," or pickleweed (e.g., Lewis 1870). The vegetation of the Walnut Creek baylands likely had similarities to the Napa River tidal marshes, with extensive tule stands along brackish channels and halophytes such as pickleweed and salt grass in areas with poorer drainage (Collins and Foin 1992, Grossinger 2012). Pannes were distributed throughout the marsh plain (Rodgers 1856, Hesse 1861, Rodgers and Chase 1866). Between Bulls Head Point and Point Edith, a wide intertidal mudflat separated the marsh plain from the deeper waters of Suisun Bay. In total, the baylands (including tidal marsh, pannes, and channels) between Bulls Head Point and Seal Bluff Landing occupied approximately 5,000 acres. To the west and east the baylands were bordered by steep hillslopes, while at the southern end of the marsh a low gradient area supported a broad freshwater-brackish transition zone (USGS [1893-4]1897, USGS 1896[1901]).

Non-tidal Wetlands

A non-tidal wetland complex, sustained by high groundwater levels, adjoined the tidal marsh on the southern end. Totaling aproximately 800 acres, these wetlands formed an ecologically complex and highly productive area which provided habitat for plant and animal species of current conservation interest such as California Tiger Salamander (CNDDB 2012). Historical evidence suggests that wetland types within this area included freshwater marsh, willow thicket, and brackish/alkali marsh and meadow. As a result of the area's flat topography, boundaries between wetland habitat types were gradual and thus challenging to define precisely. Immediately adjacent to the baylands, a perennial freshwater wetland complex was found in the area around present-day Buchanan Field Airport (Williamson 1850, USGS [1893-4]1897, USGS [1896]1901). An arm of the wetland extending downstream along the eastern side of the present-day Tesoro Refinery functioned as an overflow channel for Walnut Creek during floods. Early accounts describe a willow swamp or thicket, referred to as "Monte del Diablo" ("thicket of the devil"), within this wetland complex, though its size is not specified (Smith & Elliott [1879]1979, Viader and Cook 1960). Further south, the wetland complex graded into an alkali meadow, characterized by moderately alkaline soils, seasonal flooding, and a salt-tolerant plant community (Carpenter and Cosby 1933, 1939).

Stream Channels

Upstream of the baylands and the non-tidal wetland complex, Walnut Creek flowed along the western side of the valley (Coffee 1857, Allardt 1861, USDA 1939). A mixed riparian forest (not shown in the mapping) composed of willows, ashes, oaks, cottonwoods, alders, walnuts, and laurels lined the creek (Small 1855, Taylor 1864b, Crespí and Bolton 1927, Viader and Cook 1960). Several secondary channels branched off on the eastern side, reconnecting with the mainstem downstream or terminating in the wetland complex (Williamson 1850, McMahon and Minto 1885, USGS [1893-4]1897). Early observers noted that sections of Walnut Creek had little or no flow during the dry season: camped along the creek with the Anza Expedition in April 1776, for example, Pedro Font wrote, "[It] would be not a bad place for a settlement... if only the stream were a year-round one. But evidently it is not, as we found it having no flow and with only small pools" (Font and Brown 2011). Walnut Creek followed a meandering course through the tidal marsh to its mouth at Suisun Bay, and tides regularly overflowed the channel banks onto the marsh plain (Carpenter and Cosby 1933, 1939). In the mid-19th century, the creek was navigable as far inland as the town of Pacheco, which for a short time was an important shipping point for grain and other products (e.g., *Daily Alta California* 1860).

Numerous other streams, including Seal Creek, Mt. Diablo Creek, Galindo Creek, Pine Creek, Grayson Creek, and Hidden Valley Creek, drained into the baylands. Mt. Diablo Creek, which today drains into Seal Creek and Hastings Slough, historically flowed west through present-day Concord to connect with Walnut Creek on the southern end of the baylands (Britton & Rey 1871, Whitney 1873, McMahon and Minto 1885).

This 1895 lithograph of Diablo Valley shows "Pacheco Cr." (an early name for lower Walnut Creek) meandering through the marsh on its course to Suisun Bay. Several ships are visible navigating up the creek towards the town of Pacheco. Mt. Diablo rises in the background. (courtesy of Dean McLeod and Contra Costa County Historical Society)

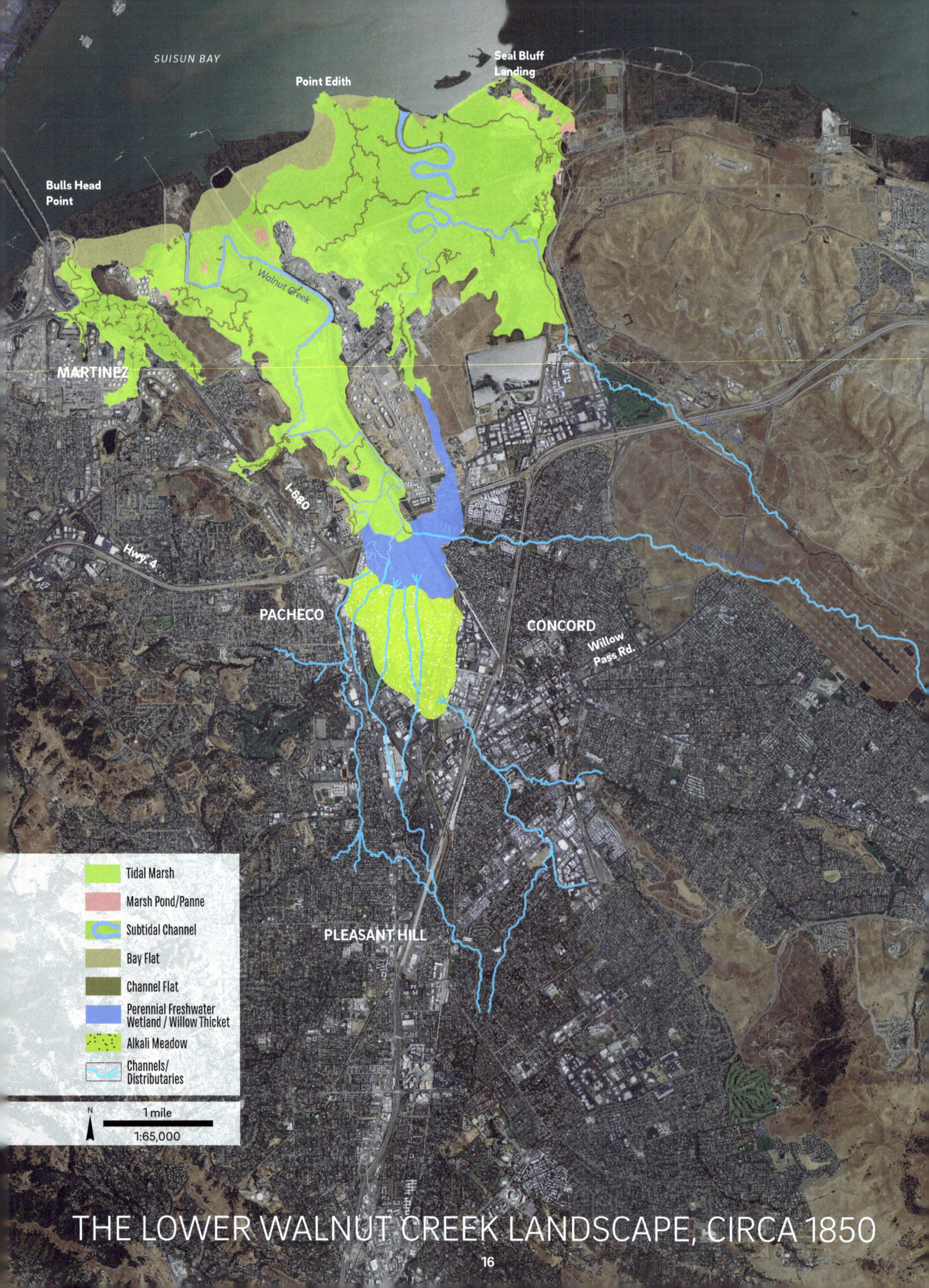

SUISUN BAY

Point Edith

Seal Bluff
Landing

Bulls Head
Point

Walnut Creek

MARTINEZ

I-680

Hwy. 4

PACHECO

CONCORD

Willow
Pass Rd.

Mt. Diablo Creek

PLEASANT HILL

Legend

- Tidal Marsh
- Marsh Pond/Panne
- Subtidal Channel
- Bay Flat
- Channel Flat
- Perennial Freshwater Wetland / Willow Thicket
- Alkali Meadow
- Channels/ Distributaries

N

1 mile

1:65,000

THE LOWER WALNUT CREEK LANDSCAPE, CIRCA 1850

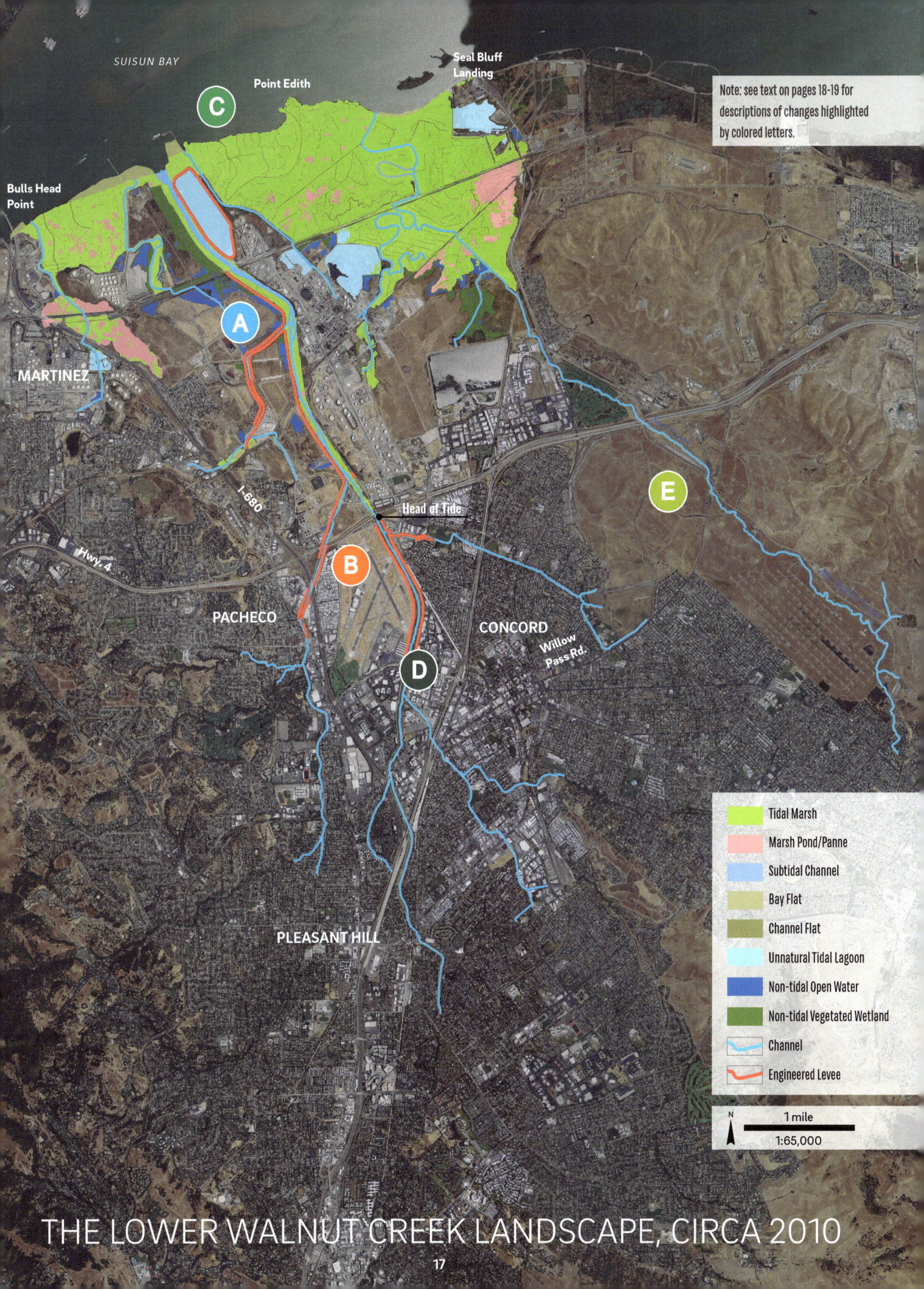

SUISUN BAY

Point Edith

Seal Bluff Landing

C

Bulls Head Point

A

MARTINEZ

I-680

Hwy. 4

B

Head of Tide

PACHECO

D

CONCORD

Willow Pass Rd.

E

PLEASANT HILL

	Tidal Marsh
	Marsh Pond/Panne
	Subtidal Channel
	Bay Flat
	Channel Flat
	Unnatural Tidal Lagoon
	Non-tidal Open Water
	Non-tidal Vegetated Wetland
	Channel
	Engineered Levee

N

1 mile

1:65,000

THE LOWER WALNUT CREEK LANDSCAPE, CIRCA 2010

LANDSCAPE CHANGE
OF LOWER WALNUT CREEK
THE LOWER WALNUT CREEK LANDSCAPE, CIRCA 2010

Note: colored letters in the text correspond to annotations on the circa 2010 map (preceding page).

The lower Walnut Creek landscape has changed dramatically over the past 150 years. Grazing and logging, which were dominant land uses in the watershed during the early to mid-1800s, likely contributed to increased rates of erosion and sediment delivery. Major roads and railroad lines were constructed through the baylands in the late 19th and early 20th centuries, reducing habitat connectivity and constricting both fluvial and tidal flows. Industrial development, urbanization, and stream channelization has greatly reduced wetland extent and altered hydrology in the lower watershed.

Loss of Tidal Wetlands

Industrial and urban development during the late 19th and 20th centuries reduced tidal wetland extent by approximately 40% between Bulls Head Point and Seal Bluff Landing. Immediately adjacent to lower Walnut Creek (excluding the Bay Flat and Unnatural Tidal Lagoon features and the wetlands west of Pacheco Marsh), approximately 85% of historical tidal wetland area has been lost, and remaining marsh areas are confined to a narrow corridor along the channel **A**. Both lateral and longitudinal connectivity in the remaining tidal wetland areas has been greatly reduced by industrial facilities, roads, and other infrastructure. In addition to the overall impact on habitat extent and quality, the loss of tidal wetlands also represents a substantial reduction in tidal prism volume, which has resulted in a decrease in channel scour capacity and contributed to sediment accumulation within the tidal portion of Walnut Creek (pages 20-23).

Loss of Non-Tidal Wetlands

The freshwater marsh, willow thicket, and seasonal alkali meadow historically found immediately south of the baylands have been eliminated by urban development **B**. Buchanan Field Airport and other developed areas occupy much of the historical footprint of this wetland

Changes in wetland extent in the lower Walnut Creek area between ca. 1850 and ca. 2010. The analysis included the full extent of tidal wetlands (both historical and modern) between Bulls Head Point and Seal Bluff Landing), the adjacent historical non-tidal wetlands, and the modern non-tidal wetlands within the historical mapping footprint.

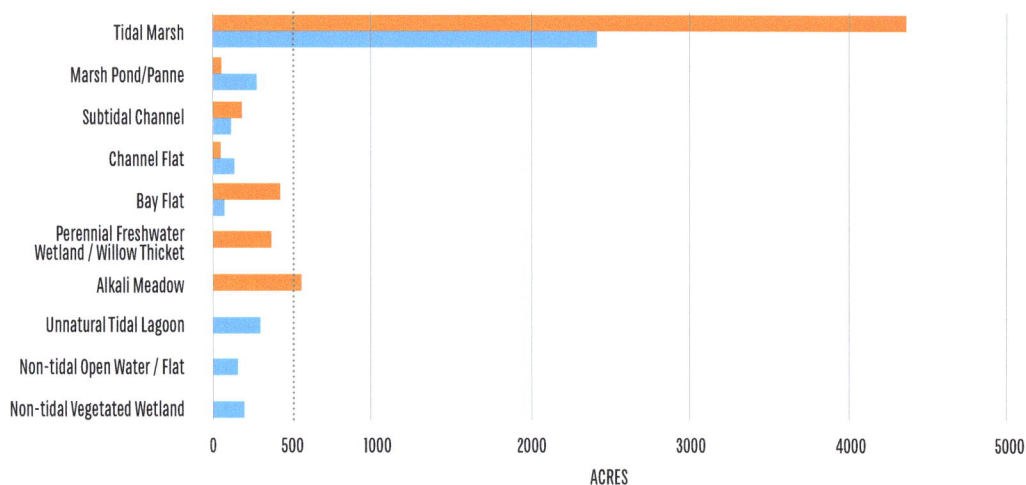

ca. 1850
ca. 2010

Distribution of Sycamores by Size at Pacheco Creek

A total of 147 live sycamores were counted at Pacheco Creek. These included 19 seedling/sucker, 38 sapling, 75 medium, and 15 large trees (Figure 5). Trees in general were most concentrated on the primary channel and inner channel corridor. The greatest density of large trees was along the inner channel corridor and/ or associated with the main tributary Harper's Creek, while the greatest density of medium trees was concentrated along the primary channel (Figure 6). Seedlings/suckers were concentrated along the primary channel and the inner channel corridor.

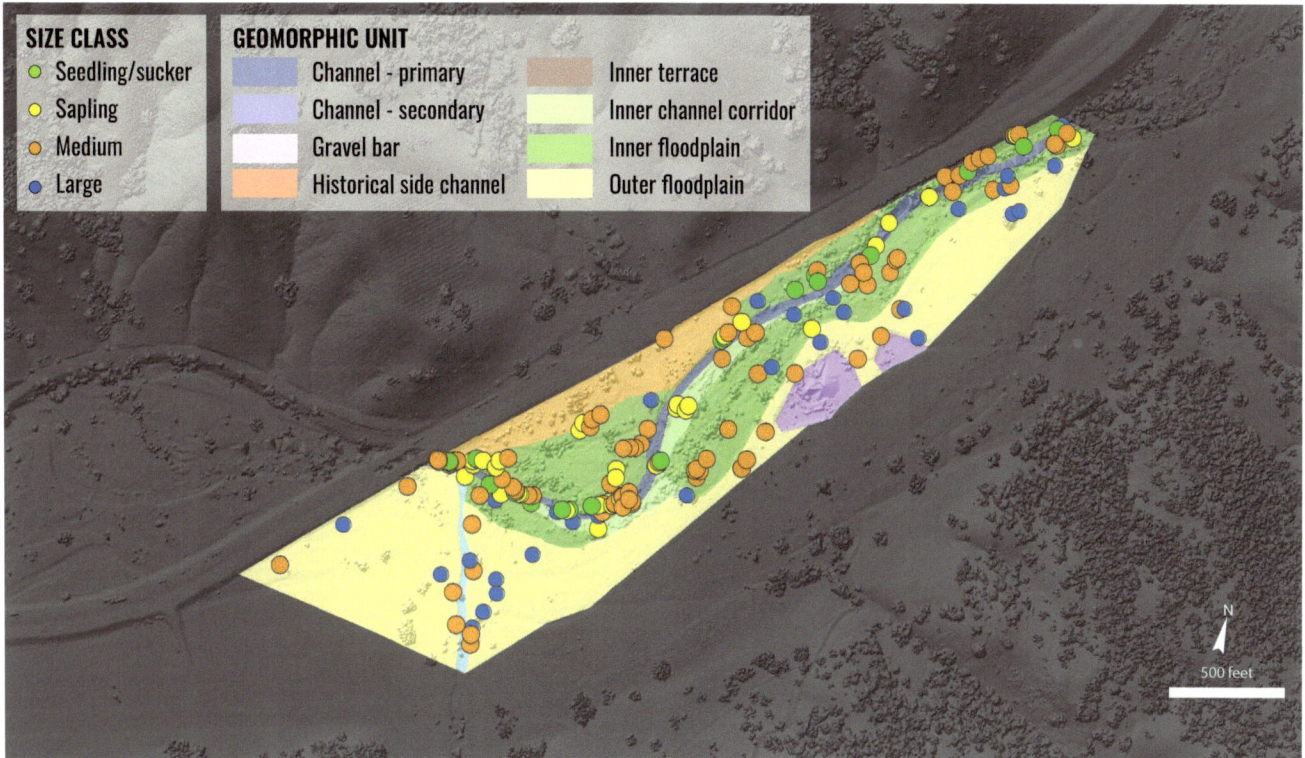

Figure 5. Map of distribution of sycamores by geomorphic zone and size class, Pacheco Creek.

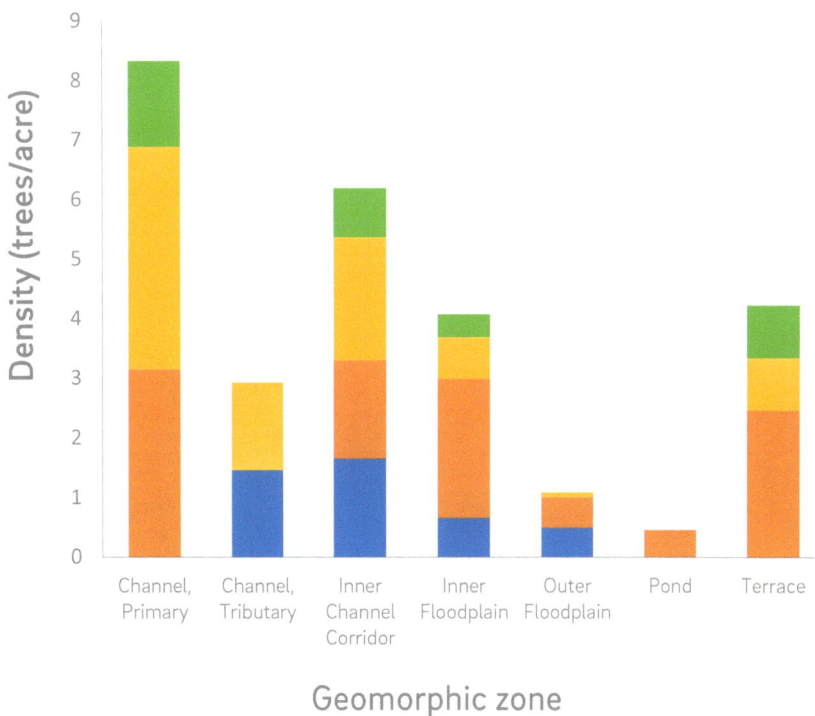

Figure 6. Density of live sycamores by geomorphic zone and size class at Pacheco Creek.

Distribution by Distance to Channel at Upper Coyote Creek

We hypothesized that larger trees would be farther from the channel and smaller trees would be closer to the channel, as this might correlate to extents of inundation. However, at Upper Coyote Creek there is no obvious, consistent pattern (Figure 7). This may be due to the changing location of the primary channel over time. As shown, medium trees are located furthest from the primary channel, while older trees on average are located an intermediate distance. Note the seedling/sucker and sapling input values are n=1 for these two size classes.

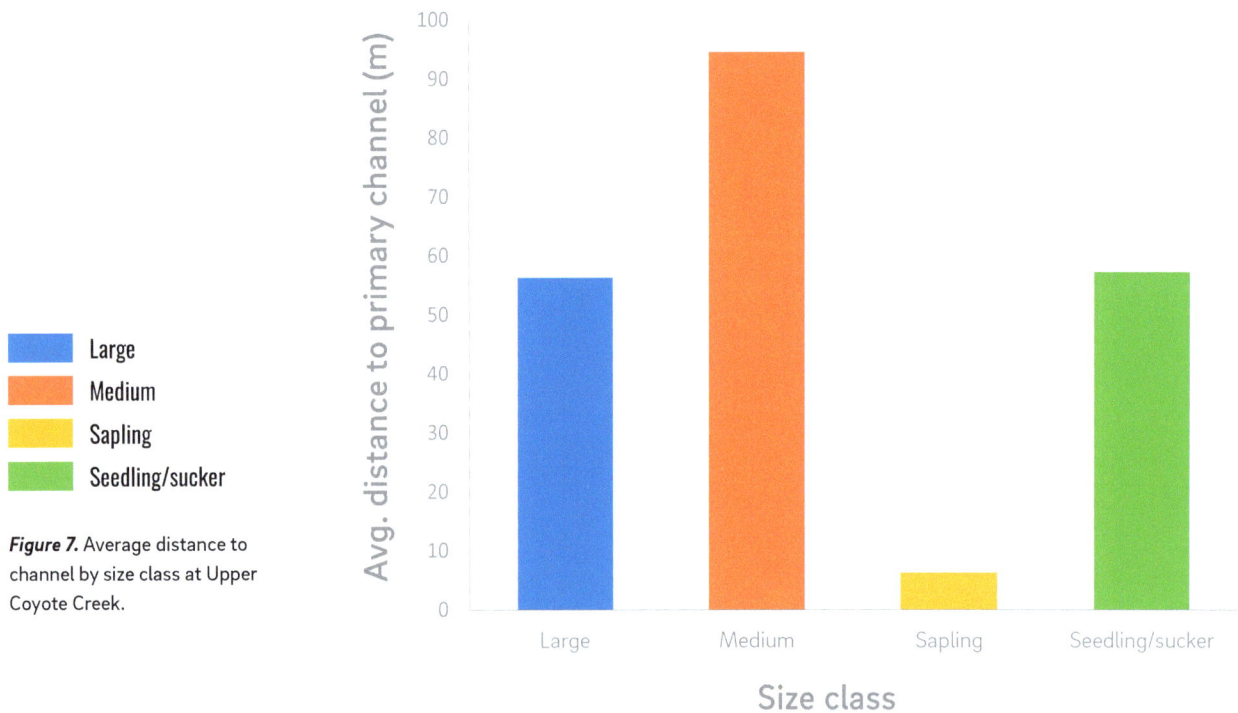

Large
Medium
Sapling
Seedling/sucker

Figure 7. Average distance to channel by size class at Upper Coyote Creek.

SYCAMORE LEAF IN FALL, PACHECO CREEK

Distribution by Distance to Channel at Pacheco Creek

As hypothesized, tree size consistently increases with distance to the primary channel at Pacheco Creek. In other words, older growth is located further from the current channel, on average and newer growth is on average closer to the primary channel (Figure 8). At Pacheco Creek there were 38 suckers, and 19 seedling/saplings recorded.

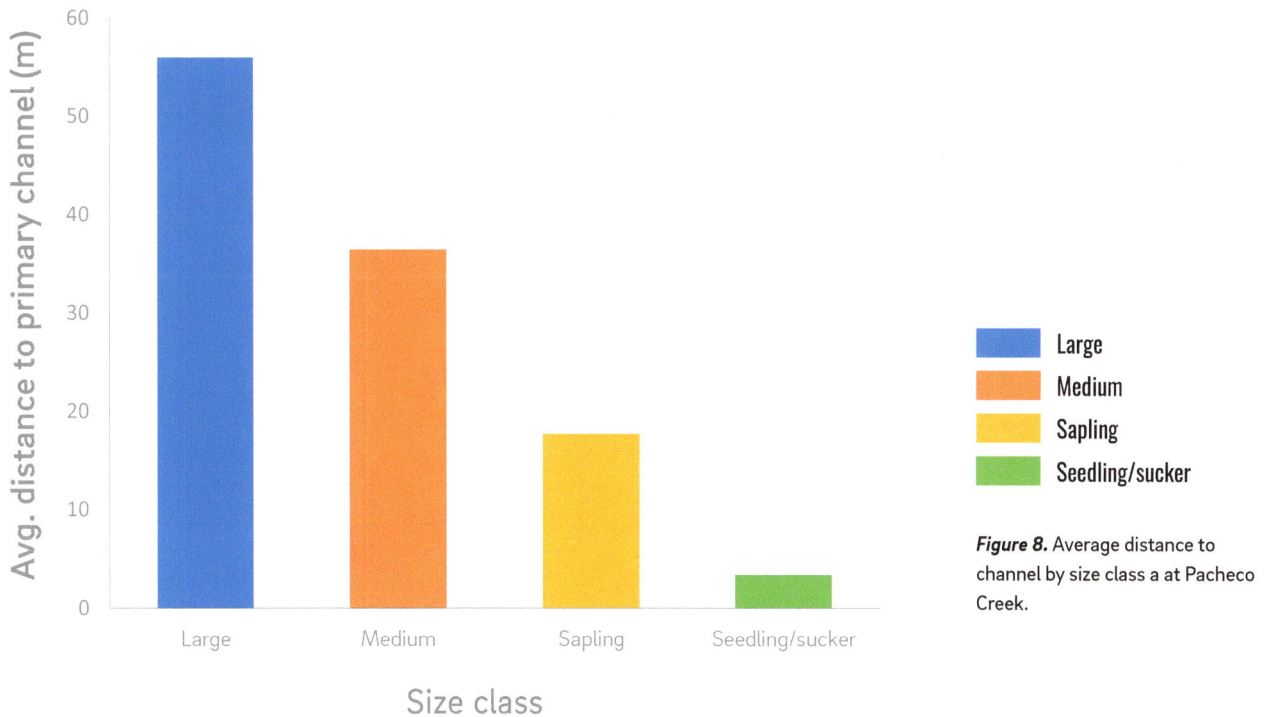

Figure 8. Average distance to channel by size class a at Pacheco Creek.

SYCAMORES AT PACHECO CREEK

SYCAMORE HEALTH PATTERNS ACROSS SITES

Evaluating general tree health can provide valuable information on overall species fitness in an area and provide a possible link to the ability of the species to successfully reproduce and sustain a viable population. There are many factors that may influence tree health, including physical damage from flood flows, insufficient groundwater levels, competition with riparian vegetation, grazing by cattle or wildlife, pests and pathogens, and anthropomorphic disturbance. This study used a combined health and vigor rating based on foliage and structure and incidence of anthracnose as general indicators of overall tree health.

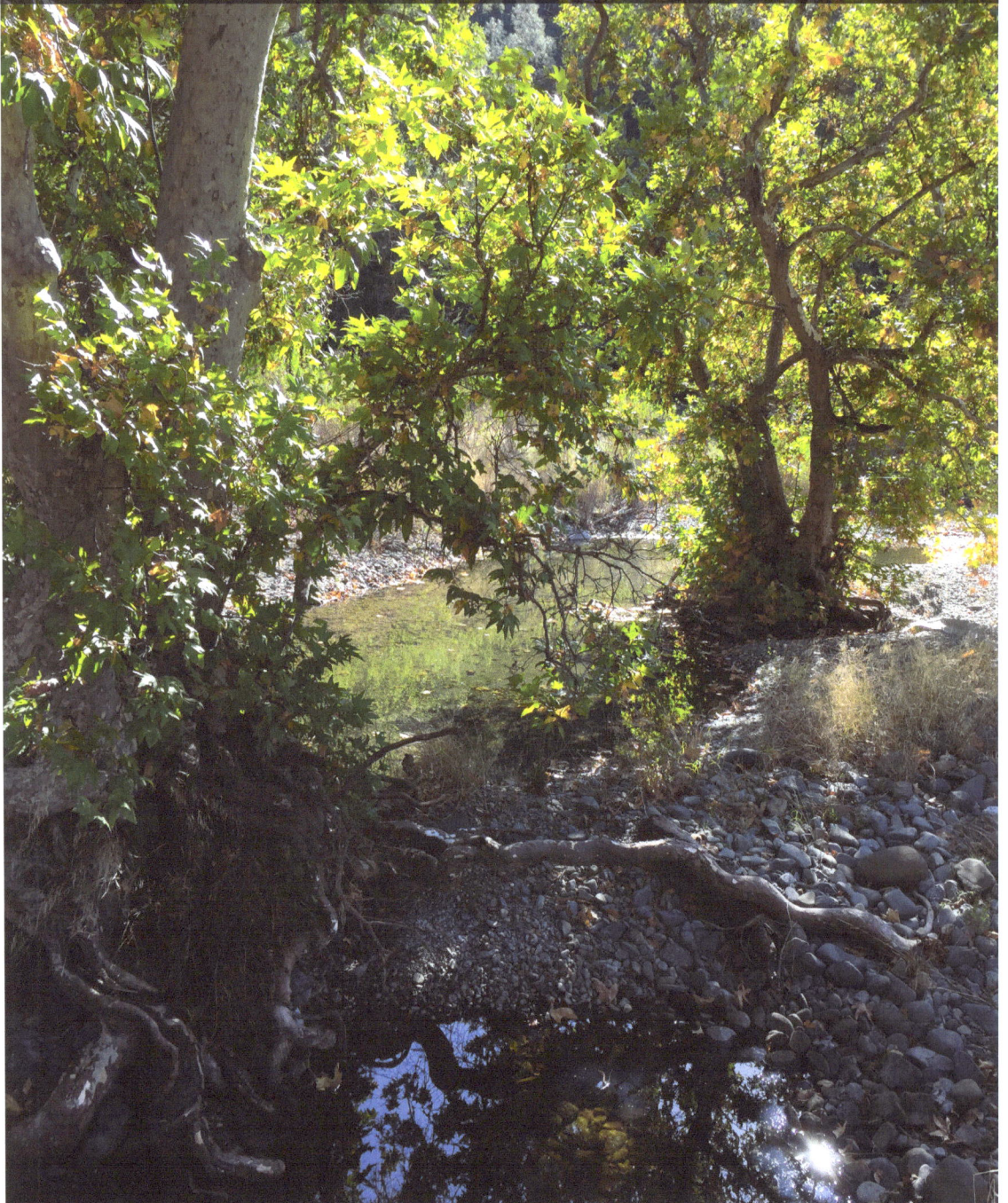

POOLS AND SYCAMORES, UPPER COYOTE CREEK

MULTISTEM TREE WITH WOODPECKER GRANARY, UPPER COYOTE CREEK

ACORN WOODPECKER, UPPER COYOTE CREEK

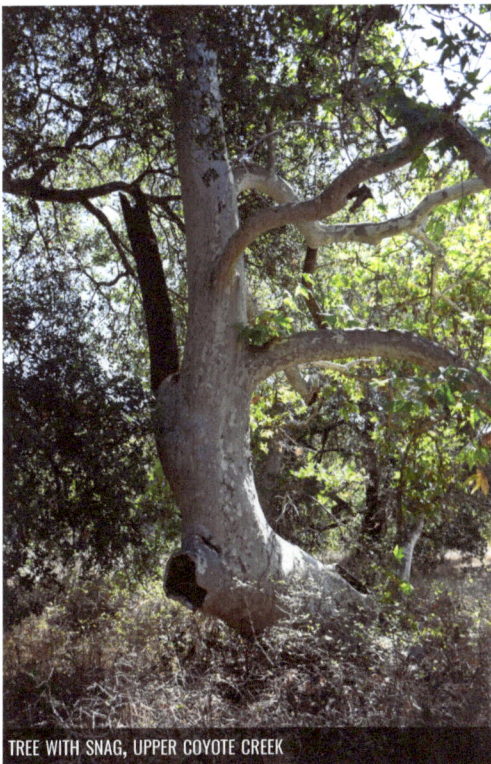
TREE WITH SNAG, UPPER COYOTE CREEK

SYCAMORE SEEDLING, PACHECO CREEK

Health Index Score by Geomorphic Zone at Upper Coyote Creek

At Upper Coyote Creek, trees are on average relatively healthy throughout the study site (Figure 9). There is no consistent, obvious gradient of tree health across geomorphic zones.

Both the highest densities and proportions of unhealthy trees are located on the historic side channel and inner channel corridor (Figure 10). Healthy trees as a proportion are somewhat evenly distributed across the site. Survivorship is remarkably high (above 98%) across size classes. Out of 304 trees, just three medium trees and one large tree was classified as dead. Survivorship was relatively consistent across geomorphic zones, with just n=4 total dead sycamores spread out in the historical side channel, primary channel and outer floodplain.

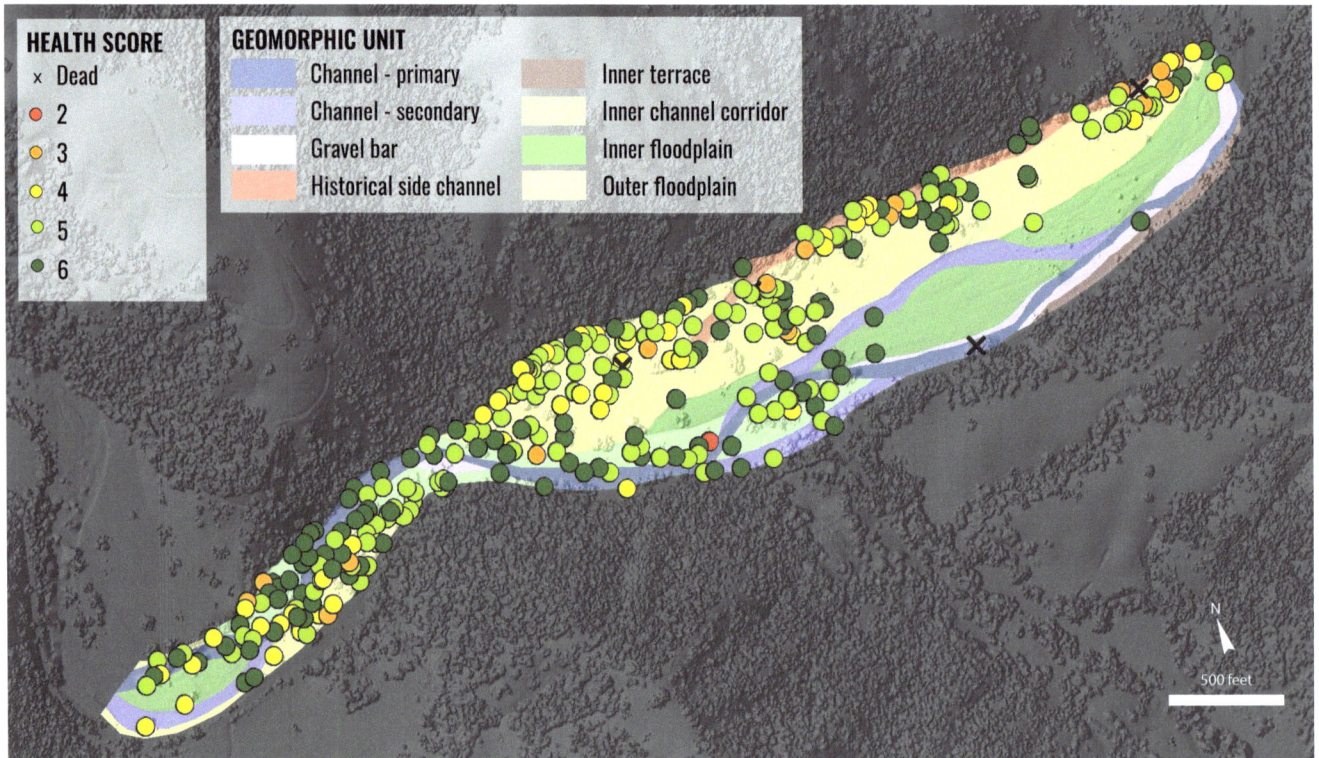

Figure 9. Distribution of trees by health index score at Upper Coyote Creek.

Figure 10. Average tree health by geomorphic zone at Upper Coyote Creek.

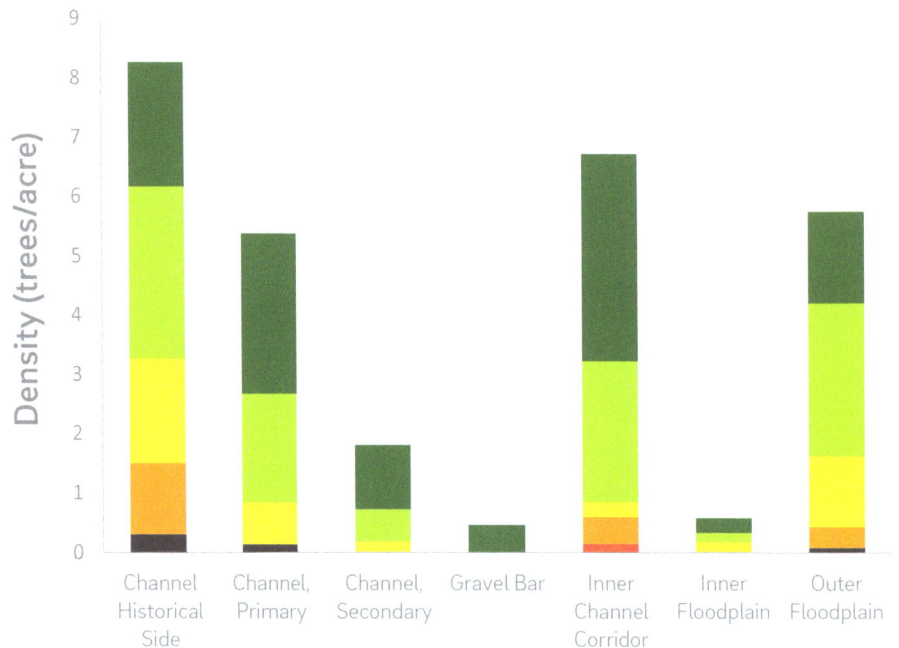

Health Index Score by Geomorphic Zone

Health Index Score by Geomorphic Zone at Pacheco Creek

At Pacheco Creek, there is no consistent gradient of tree health across geomorphic zones (Figure 11). The highest proportions of healthy trees were found in the primary channel, inner channel corridor and terrace. The highest number of dead trees were found on the outer floodplain and tributary channel (Figure 12). Close to 50% of large trees and around 15% of medium trees were classified as dead.

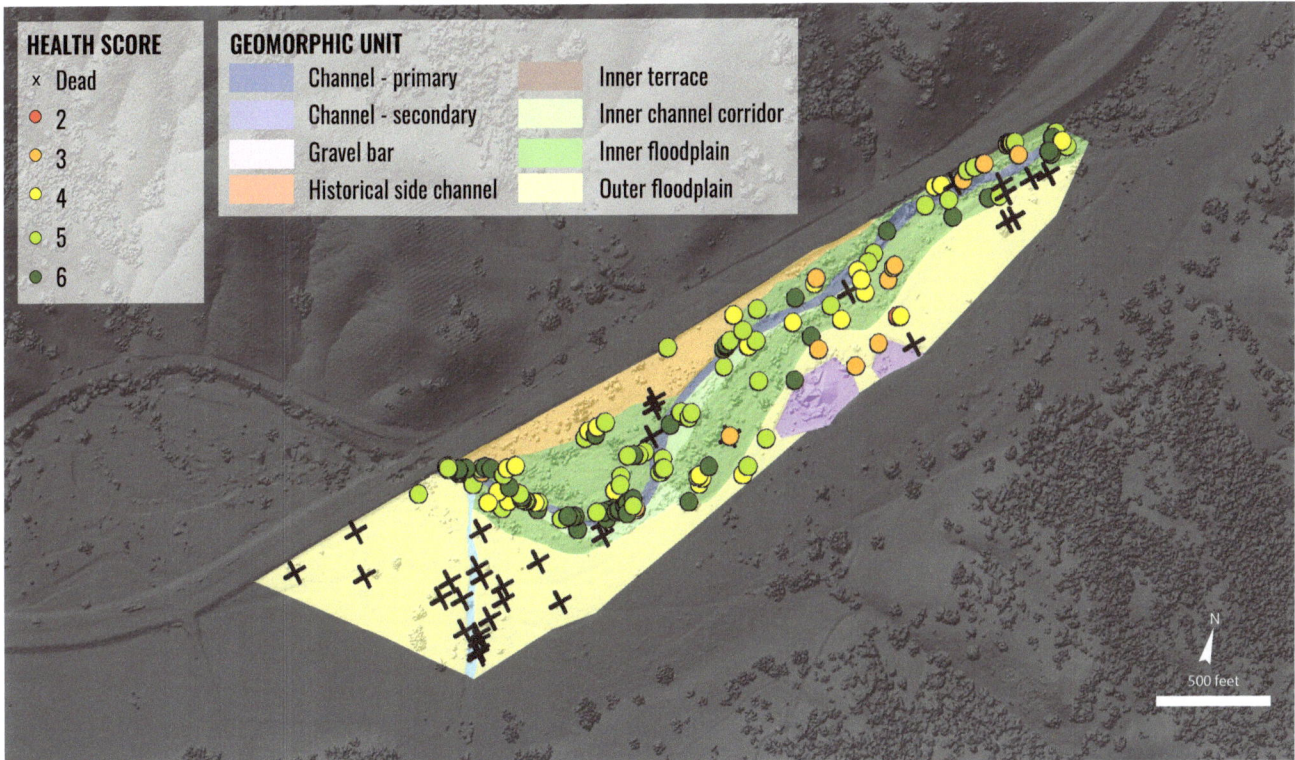

Figure 11. Distribution of trees by health index score at Pacheco Creek.

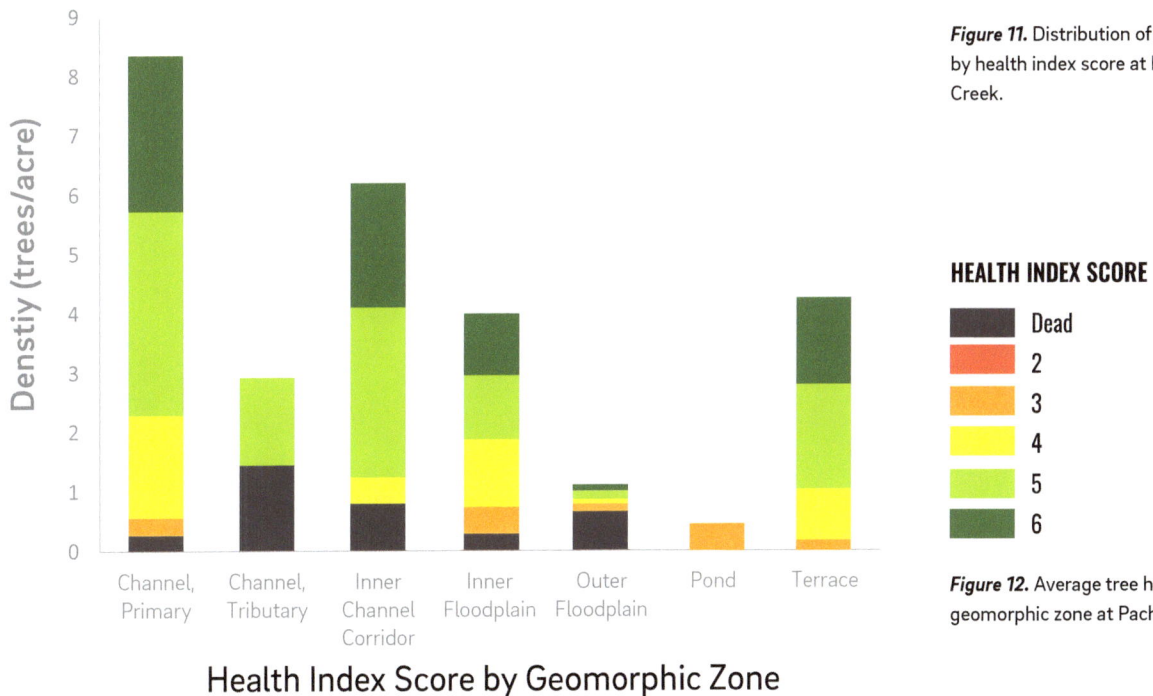

Figure 12. Average tree health by geomorphic zone at Pacheco Creek.

Health Index Score by Geomorphic Zone

29

Health Index Score by Distance to Channel at Upper Coyote Creek

At Upper Coyote Creek, live trees were on average closer to the primary channel (~70 m), while dead trees were on average >90 m from the active channel (Figure 13). The healthier trees are generally located closer to the channel. However, one should consider the small sample size for very unhealthy trees (N=1 for health score of "2").

HEALTH INDEX SCORE

- ▇ Dead
- ▇ 2
- ▇ 3
- ▇ 4
- ▇ 5
- ▇ 6

Figure 13. Relationship between health and distance to primary channel at Upper Coyote Creek.

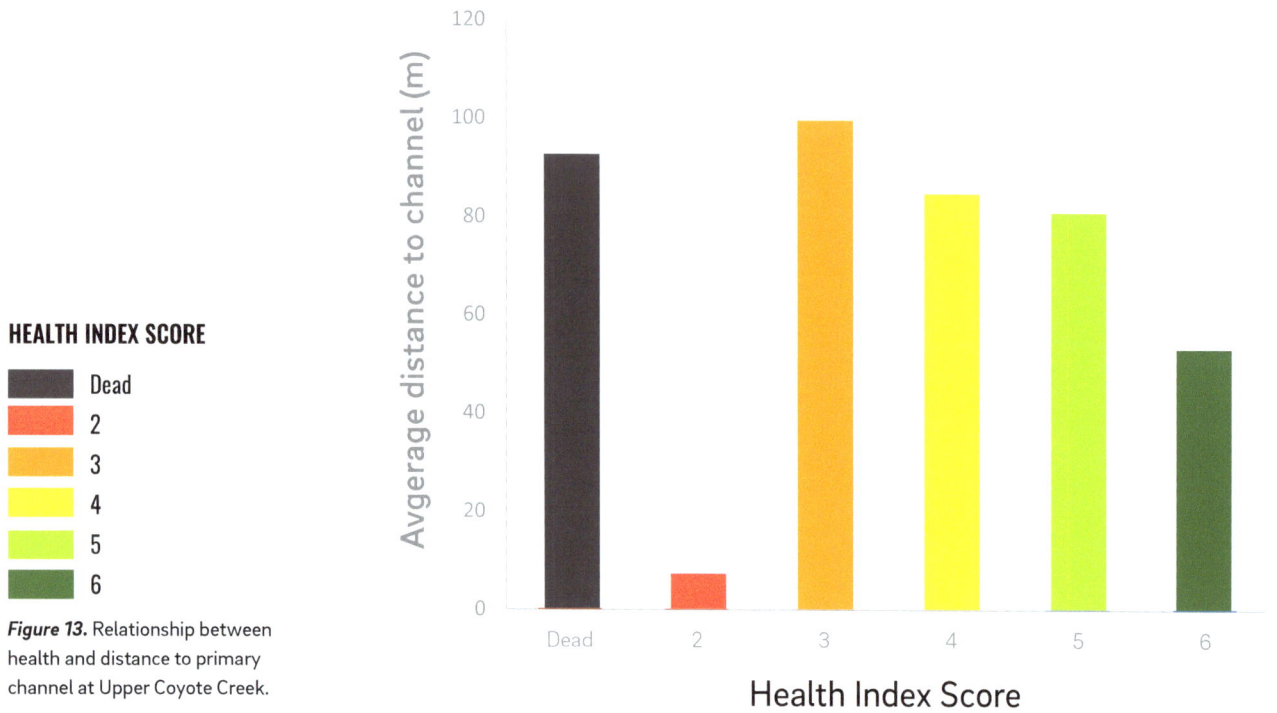

Figure 13. Relationship between health and distance to primary channel at Upper Coyote Creek.

BASE OF SYCAMORE, UPPER COYOTE CREEK

Health Index Score by Distance to Channel at Pacheco Creek

At Pacheco Creek, healthier trees were generally located closer to the channel (Figure 14). On average, dead trees were more than 90 m from the active channel, and live trees average 20 m from the active channel. Department of Fish and Wildlife staff have observed die-back of large sycamore trees at Pacheco Creek in the last several years, cause unknown (Dave Johnston, Pers. Comm.).

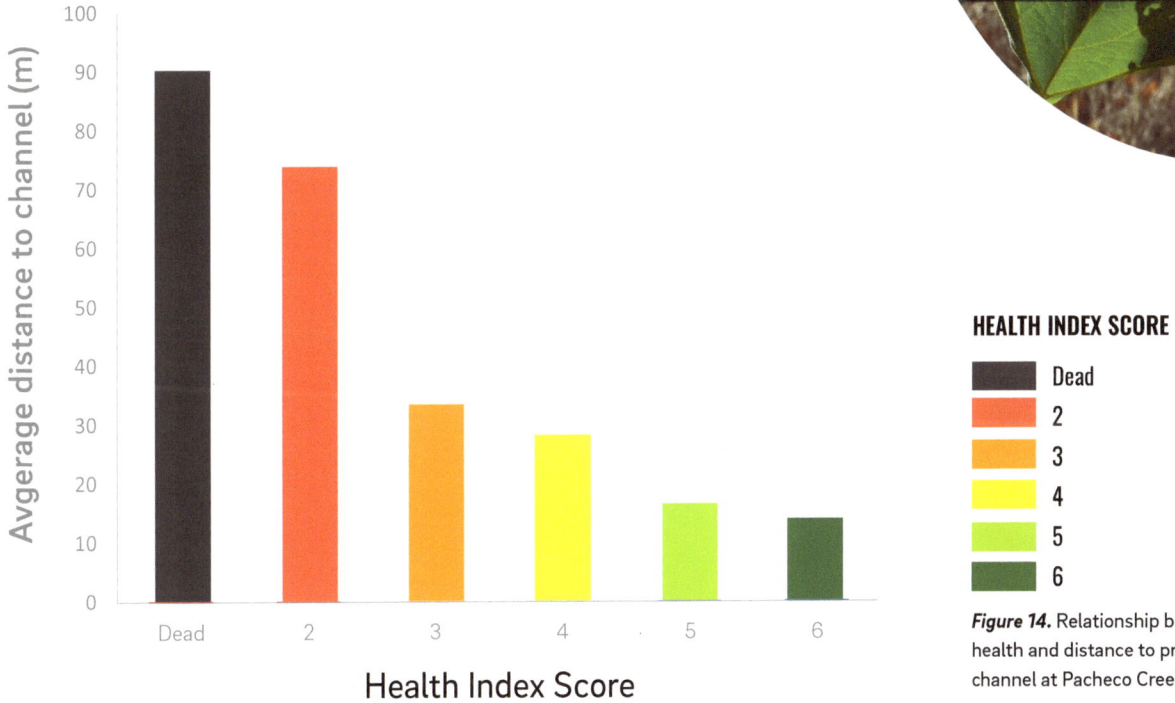

CLONAL REGENERATION, PACHECO CREEK

HEALTH INDEX SCORE

- Dead
- 2
- 3
- 4
- 5
- 6

Figure 14. Relationship between health and distance to primary channel at Pacheco Creek.

BURNED SYCAMORE WITH HOLLOWED-OUT TRUNK, PACHECO CREEK

31

Anthracnose Incidence at Upper Coyote Creek

A total of 108 trees were surveyed for anthracnose at Upper Coyote Creek. Most of the trees exhibited high foliage infestations and substantial twig die-back and/or cankers (Figure 15).

The average health and vigor of trees was medium-high and did not vary substantially by anthracnose infestation severity (Figure 16).

Anthracnose infestation severity was high across all geomorphic zones, except for one sucker/sapling on a gravel bar that exhibited minimal foliage infestation and no twig die-back.

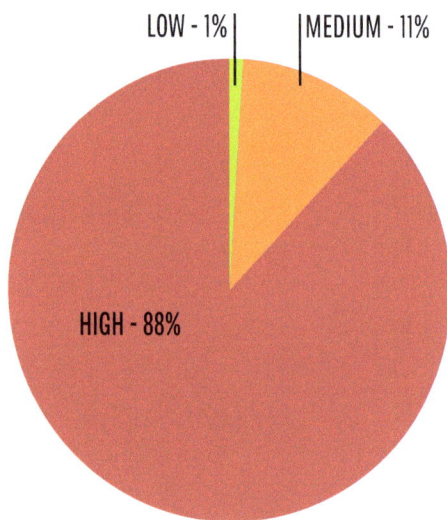

LOW - 1%   MEDIUM - 11%

HIGH - 88%

Figure 15. Mean Health and Vigor of Sycamore Trees by Anthracnose Infestation Severity Category at Upper Coyote Creek.

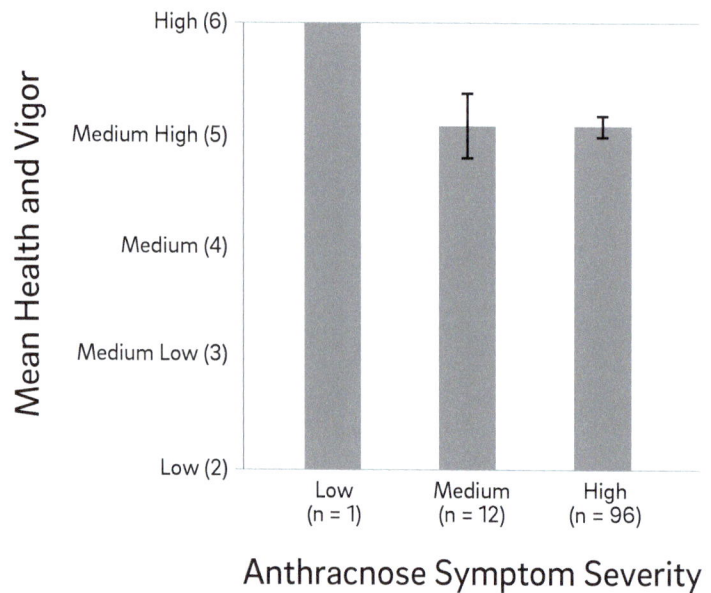

Figure 16. Mean Health and Vigor of Sycamore Trees by Anthracnose Infestation Severity Category at Upper Coyote Creek.

EVIDENCE OF ANTHRACNOSE

Anthracnose Incidence at Pacheco Creek

A total of 55 trees were surveyed at Pacheco Creek. Most of the trees exhibited low foliage infestations and little or no twig die-back (Figure 17).

The average health and vigor of trees with low anthracnose infestation severity was medium-high. The average health and vigor of trees with medium and high anthracnose infestations was medium (Figure 18). In other words, anthracnose infestation and health and vigor ratings appeared to be correlated at Pacheco, though this pattern is not strongly apparent at Coyote. Additionally, other factors such as geomorphic position are also correlated with health and vigor ratings.

Anthracnose infestation severity was low in the primary channel, tributary channel, inner channel corridor, and terrace. Anthracnose infestation severity was moderate on the inner floodplain. One tree exhibited high anthracnose infestation severity near the pond.

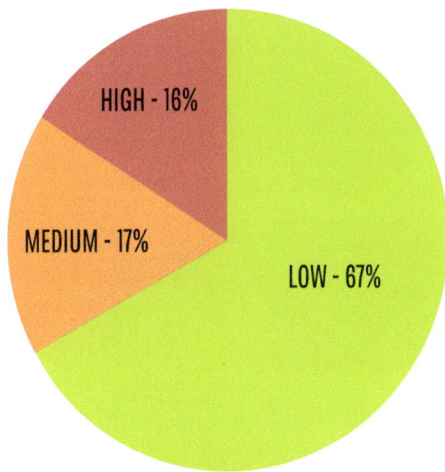

Figure 17. Percentage of Low, Medium, and High Anthracnose Infested Trees at Pacheco Creek.

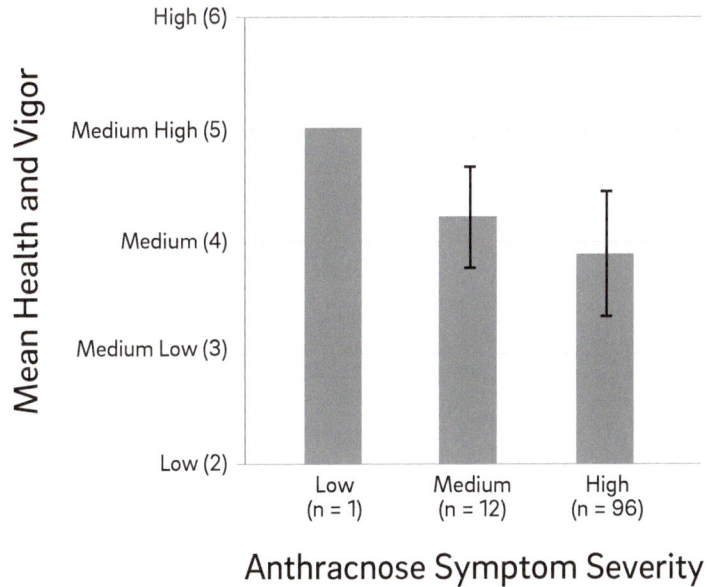

Figure 18. Mean Health and Vigor of Sycamore Trees by Anthracnose Infestation Severity Category at Pacheco Creek.

SYCAMORE REGENERATION PATTERNS ACROSS SITES

Regeneration of sycamores is generally associated with a combination of a gravel or cobble substrate (preferably recently disturbed), adequate moisture at appropriate time intervals, and adequate light. Regeneration may be limited by factors including competition from riparian vegetation, lack of freshly deposited bars and floodplains, insufficient groundwater levels, grazing from cattle or other wildlife, diseases such as anthracnose, heart rot, and climatic conditions wherein seed production and flooding do not align.

Regeneration Patterns of Sycamores at Upper Coyote Creek

Only one seedling/sucker and one sapling was found at Upper Coyote Creek, indicating that regeneration at the time of the field study was very low. Figure 19 shows the regeneration strategy observed by percentage of trees in each geomorphic zone. Some trees were sprouting, some producing seeds, some both, and some exhibited neither indicator of regeneration at the time of the field work.

Root sprouting (light green and dark green) was concentrated in the highest proportions in the inner channel corridor and the primary channel among large and medium trees. Seed production of live trees is proportionately most prolific in the inner floodplain and historical side channel. Seeds are produced almost exclusively on medium or large trees. All of the trees documented on gravel bars and in the inner channel floodplains were reproducing in some form.

Sprouts and seeds

- Sprouts, no seeds
- Seeds, no sprouts
- Seeds and sprouts
- Neither
- Insufficient data

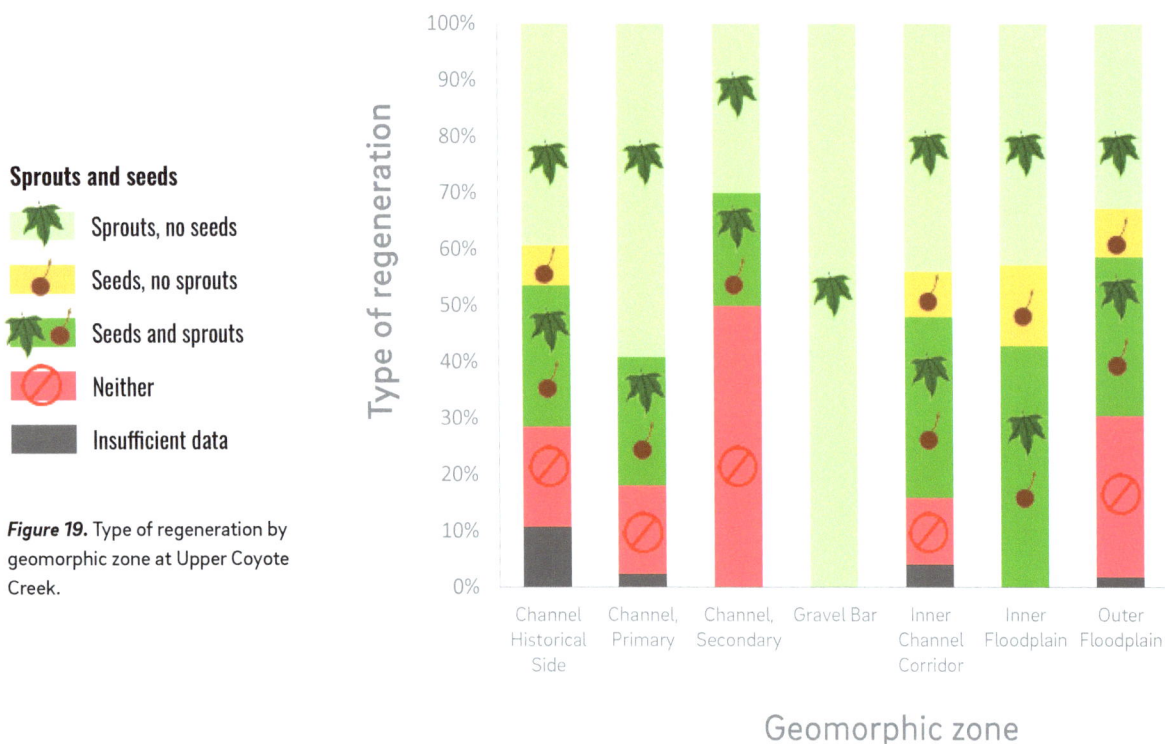

Figure 19. Type of regeneration by geomorphic zone at Upper Coyote Creek.

Regeneration Patterns of Sycamores at Pacheco Creek

Root sprouting was concentrated in the highest proportions in the inner channel corridor and inner floodplain. Seed production of live trees was proportionately most prolific in the inner floodplain and inner channel corridor. Seeds were produced almost exclusively on large or medium trees. Almost 40% of the trees on the primary channel, and 50% on the terrace, were not producing seeds or sprouts. Figure 20 shows the relative proportions of trees using different reproduction strategies.

At Pacheco Creek, 19 seedling/suckers and 38 saplings were found. Regeneration was highest in the primary channel and inner channel corridor, while more recent seedlings were also concentrated along the terrace. Cobble was a dominant substrate for younger trees across habitat types, with grassy and woody surfaces also utilized. Regeneration followed a trend of increased growth the closer a tree is to the primary channel (Figure 21).

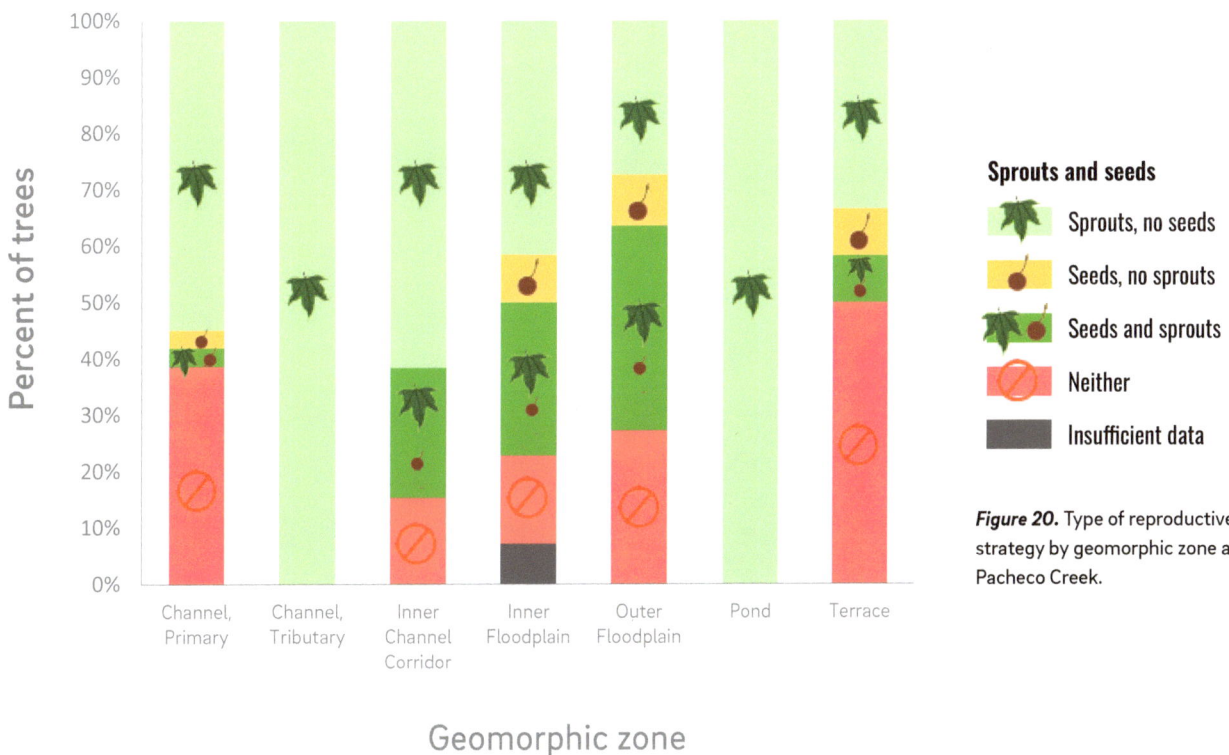

SYCAMORES RESPROUTING AT UPPER COYOTE CREEK

Sprouts and seeds

- Sprouts, no seeds
- Seeds, no sprouts
- Seeds and sprouts
- Neither
- Insufficient data

Figure 20. Type of reproductive strategy by geomorphic zone at Pacheco Creek.

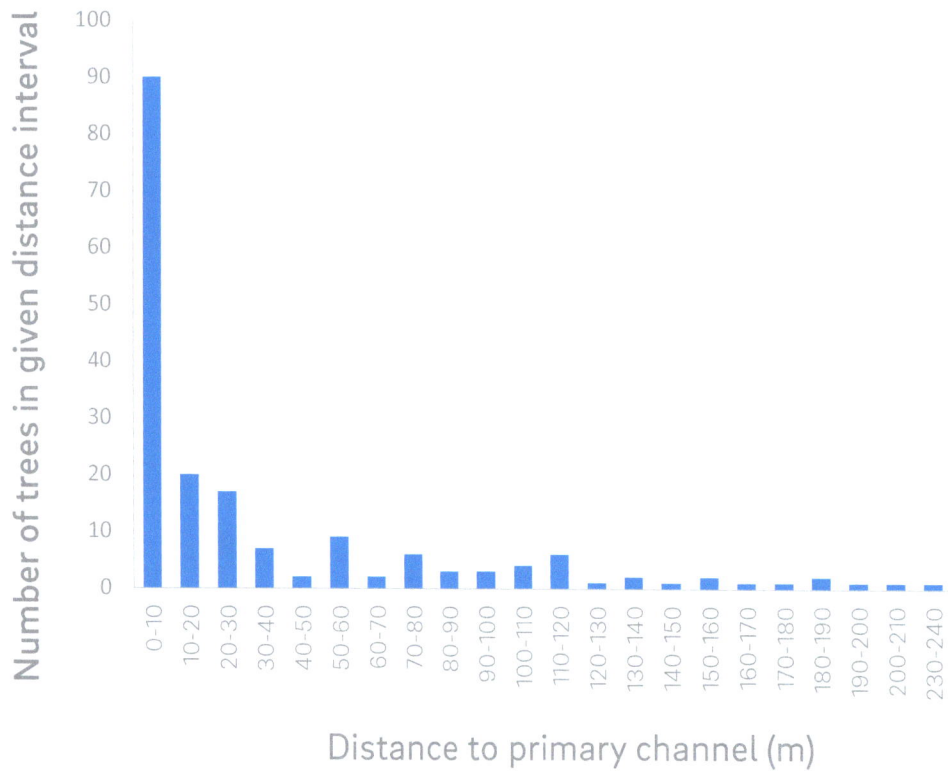

Figure 21. Regeneration of seedling/sucker & sapling by distance to channel at Pacheco Creek.

Regeneration Patterns: Species associations at Upper Coyote Creek

The species associated with regenerating sycamore trees (seedlings/suckers and saplings) at Upper Coyote Creek are described in Table 5. Most of these species occur widely in the region and are notably drought-tolerant. Only two species with a wetland indicator status of 'Facultative' or wetter were observed to be associated with regenerating sycamore trees. The percent of sycamore seedlings/suckers and saplings listed in Table 5 does not necessarily indicate actual associations between plant species and regenerating sycamores because the small sample size of two seedling/sucker sycamore trees has very low statistical power. Nonetheless, the list of species does give some indication of which dominant species are present around each of the two seedling/sucker trees at Upper Coyote Creek.

Table 5.   Most Common Plant Species Associated with Regenerating Sycamore Trees (Seedlings/Suckers and Saplings) at Upper Coyote Creek

Scientific Name	Common Name	Origin	Growth Form	Wetland Indicator Status	Percent of Sycamore Seedlings/suckers and Saplings (n = 2) (%)
Avena fatua	Wild oats	Nonnative	Grass	UPL	50.0
Bromus diandrus	Ripgut brome	Nonnative	Grass	UPL	100.0
Baccharis pilularis	Coyote brush	Native	Shrub	UPL	50.0
Baccharis salicifolia	Mulefat	Native	Shrub	FAC	50.0
Brassica nigra	Black mustard	Nonnative	Herb	UPL	50.0
Juncus patens	Spreading rush	Native	Herb	FACW	50.0
Toxicodendron diversilobum	Poison oak	Native	Shrub	FACU	50.0

Regeneration Patterns: Species associations at Pacheco Creek

The species associated with the 58 regenerating sycamore trees (seedlings/suckers and saplings) at Pacheco Creek are described in Table 6. Many of these species had a wetland indicator status of facultative or wetter, which may indicate that regeneration of sycamore trees occurred in locations with wetland hydrology.

Table 6. Most Common Plant Species Associated with Regenerating Sycamore Trees (Seedlings/Suckers and Saplings) at Pacheco Creek

Scientific Name	Common Name	Origin	Growth Form	Wetland Indicator Status	Percent of Sycamore Seedlings/suckers and Saplings (n = 58) (%)
Baccharis pilularis	Coyote brush	Native	Shrub	UPL	19.0
Brickellia californica	Brickel bush	Native	Herb	FACU	6.9
Juncus mexicanus	Mexican rush	Native	Herb	FACW	22.4
Juncus xiphioides	Iris leaved rush	Native	Herb	OBL	10.3
Persicaria amphibia	Water smartweed	Native	Herb	OBL	6.9
Rumex salicifolia	Willow dock	Native	Herb	FACW	34.5
Salix lasiolepis	Arroyo willow	Native	Shrub	FACW	6.9
Sambucus nigra spp. caerulea	Blue elderberry	Native	Shrub	FAC	15.5
Solanum sp.	Nightshade	Unknown	Unknown	Unknown	8.6
Toxicodendron diversilobum	Poison oak	Native	Shrub	FACU	44.8

Regeneration Patterns: Substrate type by size class at Upper Coyote Creek

Approximately 62.1% of the sycamores were located on vegetated substrate dominated by grasses, including one seedling/sucker, 90 medium trees, and 89 large trees (Table 7). Nineteen percent of the sycamores were located on cobbles, including one sapling, 14 medium trees, and 40 large trees. Approximately 13.4% of the sycamores were located on vegetated substrate dominated by woody vegetation, including 14 medium trees and 25 large trees. The remaining 5.5% of the sycamores were located on gravels, bare soil, or other substrates.

Table 7. Percentage and Number of Sycamores within Substrate Type by Size Class at Upper Coyote Creek

Substrate	Percentage and Number (n) of Live Sycamores by Size Class				
	Seedling/sucker	Sapling	Medium	Large	Total
Cobbles	0.0 (0)	1.8 (1)	25.5 (14)	72.7 (40)	19.0 (55)
Gravels	0.0 (0)	0.0 (0)	14.3 (1)	85.7 (6)	2.4 (7)
Soil	0.0 (0)	0.0 (0)	100.0 (3)	0.0 (0)	2.1 (6)
Vegetated Surface (grasses)	0.6 (1)	0.0 (0)	50.0 (90)	49.4 (89)	1.0 (3)
Vegetated Surface (woody)	0.0 (0)	0.0 (0)	35.9 (14)	64.1 (25)	62.1 (180)
Other	0.0 (0)	0.0 (0)	33.3 (2)	66.7 (4)	13.4 (39)

Regeneration Patterns: Substrate type by size class at Pacheco Creek

At Pacheco Creek, approximately 39.2% of the sycamores were located on vegetated substrate dominated by grasses, including two seedling/sucker, 10 saplings, 36 medium trees, and eight large trees (Table 8). Twenty-eight percent of the sycamores were located on cobbles, including nine seedlings/suckers, 17 saplings, 11 medium trees, and three large trees. Approximately 16.1% of the sycamores were located on vegetated substrate dominated by woody vegetation, including three seedlings/suckers, nine saplings, nine medium trees and two large trees. The remaining 16.8% of the sycamores were located on gravels, bare soil, or other substrates.

Table 8. Percentage and Number of Sycamores within each Substrate Type by Size Class at Pacheco Creek

Substrate	Percentage and Number (n) of Live Sycamores by Size Class				
	Seedling/sucker	Sapling	Medium	Large	Total
Cobbles	22.5 (9)	42.5 (17)	27.5 (11)	7.5 (3)	28.0 (40)
Gravels	0.0 (0)	0.0 (0)	100.0 (4)	0.0 (0)	2.8 (4)
Soil	30.0 (3)	0.0 (0)	50.0 (5)	20.0 (2)	7.0 (10)
Vegetated Surface (grasses)	3.6 (2)	17.9 (10)	64.3 (36)	14.3 (8)	39.2 (56)
Vegetated Surface (woody)	13.0 (3)	39.1 (9)	39.1 (9)	8.7 (2)	16.1 (23)
Other	20.0 (2)	0.0 (0)	80.0 (8)	0.0 (0)	7.0 (10)

RESULTS OF HYDROLOGIC ASSESSMENT AND TREE CORING

Floods that carry seeds, deliver sediment, and inundate floodplains may play a key role in sycamore regeneration. Over the past 180 years, changes to flow and sediment dynamics from dams and the removal of floodplains from the influence of regular flooding have greatly impacted the distribution and regeneration of sycamore-alluvial woodlands (Keeler-Wolf et al. 1996). Specific flood timing and frequencies links to sycamore regeneration are currently not well understood.

In order to improve management approaches that support SAW, we need to understand the specific flood requirements for large-scale regeneration. This understanding can help identify opportunities to improve regeneration of SAW sites through modifications to reservoir operations and flow hydrographs.

In an effort to understand the relationship of flood history and sycamore regeneration at the two sites in this study, we used hydrologic records and geomorphic information at both study sites with a simple hydraulic model to estimate flood depths and extents for several flood events. We then cored several trees at each site to correlate large flood event magnitude with sycamore regeneration (for a full discussion of methods, see pages 20-21).

Two primary questions guided this phase of the research:

(1) What are the estimated inundation depths and extents associated with different recurrence-interval discharges?

(2) How do specific historical flood events relate to the location of existing sycamore stands along the creek?

Hydrologic/Hydraulic Results

We calculated flood frequency curves using gage data from Coyote Creek near Gilroy (USGS gage 11169800), which is 1.5 km downstream of the study site, and Pacheco Creek near Dunneville (USGS gage 11153000), which is 6 km downstream of the Pacheco Creek study site (Tables 9 and 10). Using the calculated flood frequency curves and the hydraulic model on one cross section per site, we estimated the inundation height for a range of return intervals at both sites.

Table 9. Flood frequency analysis for Upper Coyote Creek

Return Period (year)	Estimated Discharge (cfs)
1.5	1800
2	3200
5	7200
10	9800
50	15000
100	16000

Table 10. Flood frequency analysis for Pacheco Creek

Return Period (year)	Estimated Discharge (cfs)
1.5	700
2	1700
5	6400
10	11000
50	24000
100	29000

Figure 22. (below) LiDAR-derived cross section on Upper Coyote Creek site with modeled flood inundation depths.

Figure 23. (below right) Flooded primary channel and gravel bar at Upper Coyote Creek site, January 24, 2016. Dotted red line indicates extent of flooding. Photo by Dan Stephens HTH.

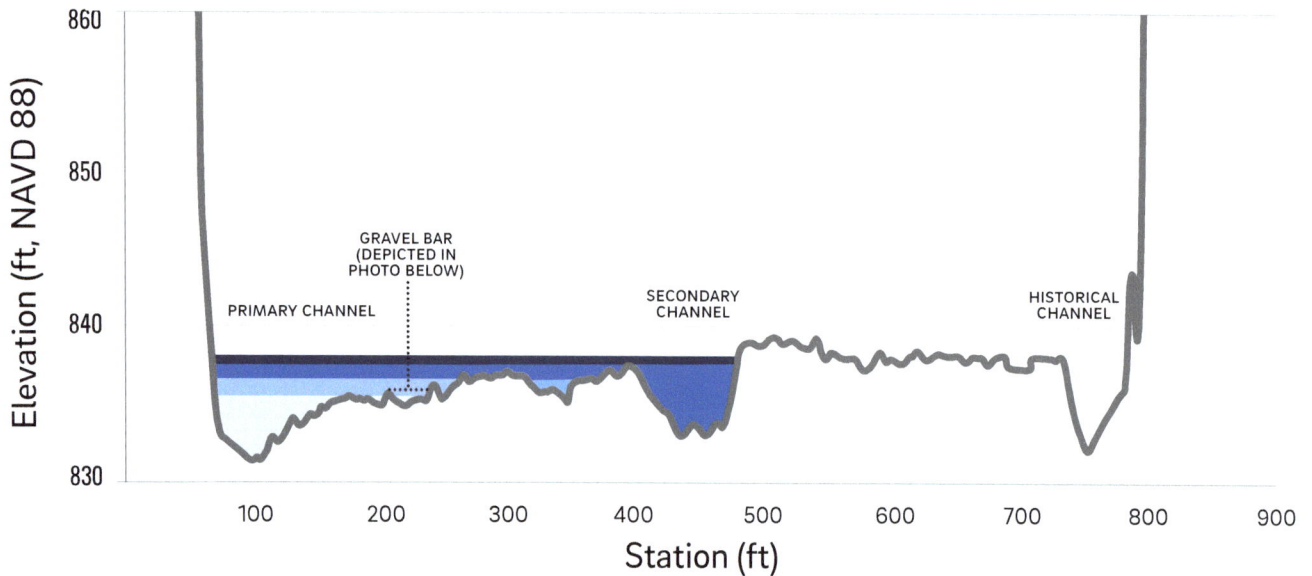

Upper Coyote Creek Initial results from the hydraulic mode model, using a LiDAR-derived cross-section and slope estimate (0.0046), pebble count data from the primary channel (D50= 100 mm), roughness values estimated from the field, and other sources (n=0.045, (Barnes, 1967), suggest that the two-year flood event (Q_2 event) fills the primary channel, and inundates part of inner floodplain. High flow indictors observed after the January 19-20, 2016 high flow event, which was estimated be a 1.5-year flood, corroborate this finding. We also found that the Q_{10} and higher flood events inundate the primary and secondary channel, as well as the gravel bar and inner floodplain (Figure 22). These results also suggest that it takes a Q_{50} event to flood the entire valley. Further study is needed to understand when the "historical channel" was active (presumably as the primary channel), and when it presumably anastamosed, or jumped, to the current primary channel location.

FLOOD EVENTS

- ~~~ Channel profile
- Flood event: Jan 1, 2016
- Flood event: Q_2
- Flood event: Q_{10}
- Flood event: Q_{50}

Location of Coyote Creek cross section

Pacheco Creek According to the hydraulic model, using a LiDAR-derived cross-section and slope estimate (0.0095), pebble count data from the primary channel (D50=45 mm), and roughness values estimated from the field, and other sources (n=0.056; Barnes, 1967), as well as gaging data adjusted by a 12% reduction in drainage area between the gage and the site, the Q_2 flow stays within the primary channel while the Q_5 and Q_{10} floods fill the primary channel and inundate the inner channel corridor. Between Q_{10} and Q_{50} floods, the inner floodplain becomes inundated. These findings suggest that the mature sycamore stands on Pacheco Creek's outer floodplain (which were not mapped as part of this study) were established during at least a 25-year flood flow, which occurred in 1956 and 1969.

It is important to note that the analyses presented here are estimates based on the best available data and a simple hydraulic analysis at two cross sections. As such, they should be considered a first approximation of flooding depths and extents. A next step to refine these results could be done by conducting a more detailed hydraulic modeling analysis that includes other factors such as changes to channel dimensions during large flood events.

Figure 24. LiDAR-derived cross section on Pacheco Creek site with modeled flood inundation depths.

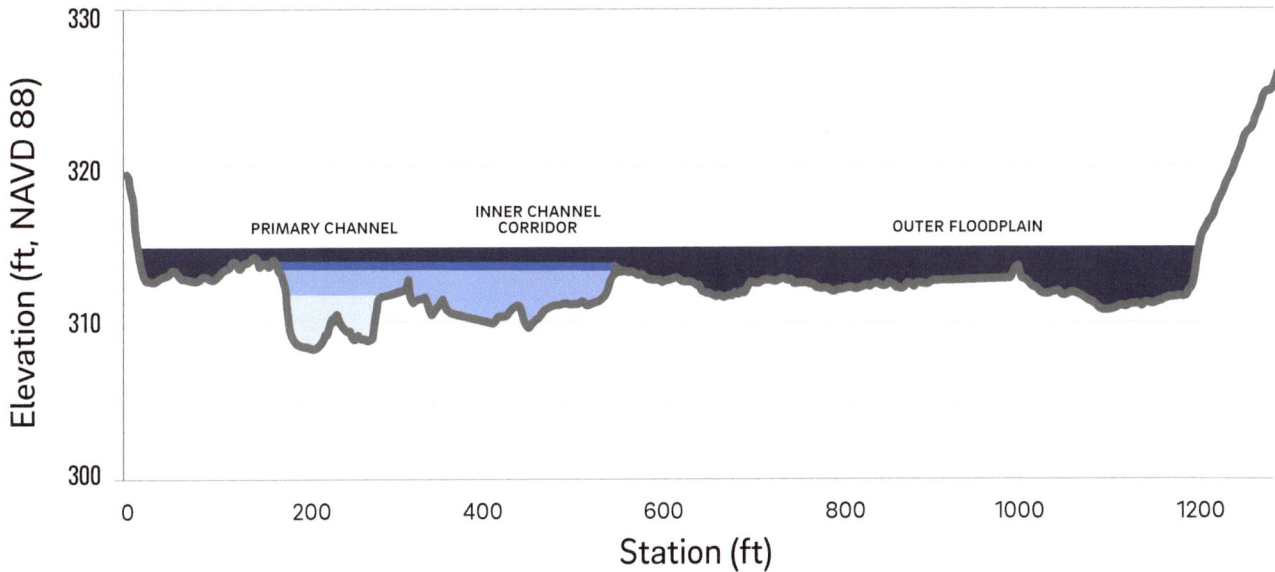

FLOOD EVENTS

— Channel profile
Flood event: Q_2
Flood event: Q_5
Flood event: Q_{10}
Flood event: Q_{50}

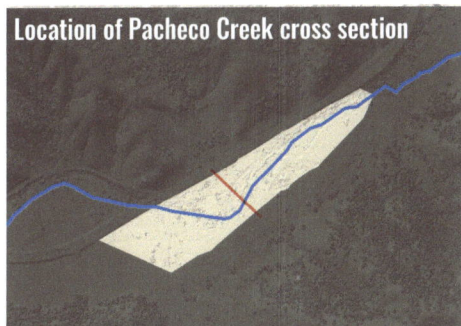

Location of Pacheco Creek cross section

HIGH FLOWS ON PACHECO CREEK (2017)

Tree Coring Results

We cored individual sycamores at the two sites to help determine the relationship between specific flood events and sycamore recruitment in the study area. In October 2016, SFEI staff cored five trees at Upper Coyote Creek and three trees at Pacheco Creek. Initial tree selection occurred before the field effort using criteria described on page 20. While in the field, we discovered that several pre-selected trees did not meet our study criteria and we were forced to find alternates. Several trees were found to have heart rot and complete cores were not possible to obtain.

Upper Coyote Creek At Coyote Creek, we cored three trees in the historical channel, one tree in the secondary channel, and one tree in the primary channel (Figure 25). The results are shown in Table 11.

Figure 25. Locations of trees cored on Upper Coyote Creek. (NAIP 2014)

Table 11. Results of trees ring analysis on Upper Coyote Creek

Site	Tree ID	Length of core (cm)	Age (est)	Error (+/-)	% Error	Heart Rot	Minimum year est.
Coyote	65	12.1	35	4	11.4	Yes	na
Coyote*	41	12.6	32	6	18.8	Yes	na
Coyote*	41	14.9	42	5	11.9	Yes	na
Coyote	9	18.2	67	6	9.0	Yes	na
Coyote	27	44.7	99	15	15.2	No	1917
Coyote	55	8.8	33	3	9.1	Yes	na

*two cores were taken for tree 41, and both had heart rot.

Because of the presence of heart rot, we were unable to attain a complete core for four of the five trees, preventing the determination of absolute age in those trees. This led to further questions about the health of older sycamore trees at Upper Coyote. Tree 27, for which it was possible to obtain an age, was not rotten at the center. Tree 27 was very straight and did not have the snags, snarls and shape of other nearby sycamores, suggesting it might be a hybrid (Figure 26). Whitlock (2003) expressed concern that sycamore hybrids may be less susceptible to diseases that cause deadwood and cavities within trunks that provide important habitats for several riparian species.

The age of Tree 27, which is located along the secondary channel, was estimated to be 99 years old with a 15% error margin. This would date the tree to around 1917 plus or minus 15 years, which is approximately three decades before the gaging records start at Upper Coyote Creek. However, using a regression analysis with gage data from a nearby gage on Alameda Creek at Niles Canyon, we were able to estimate the Coyote Creek peak flood discharge for this time period (Figure 27). From this analysis, we estimate that large storm events took place in 1911, 1919, and 1922. These were in the vicinity of 25-year storm events, according to the flood frequency analysis. Our simple hydraulic analysis shows this size storm would have inundated both the main and secondary channels, and could possibly have provided the disturbance needed to support the establishment of Tree 27 and others. This size storm also occurred at the site in 1995 and 1998, yet trees in the range of 20-22 years old are not found at the site. On January 10, 2017, Upper Coyote Creek flowed at over 10,000 cfs, which is approximately a 25-year event in this system, and similar in size to the 1919, 1922, 1995, and 1998 events. It remains to be seen what, if any, regeneration stems from this event. However, if past floods are an indication, it is possible that a 25-year event is a minimum threshold for vegetation "resetting" at this site.

ROOT SPROUTING AND HOLLOW STUMP AT UPPER COYOTE CREEK

SYCAMORE ' TREE 27', PROBABLY OVER 100 YEARS OLD. COYOTE CREEK

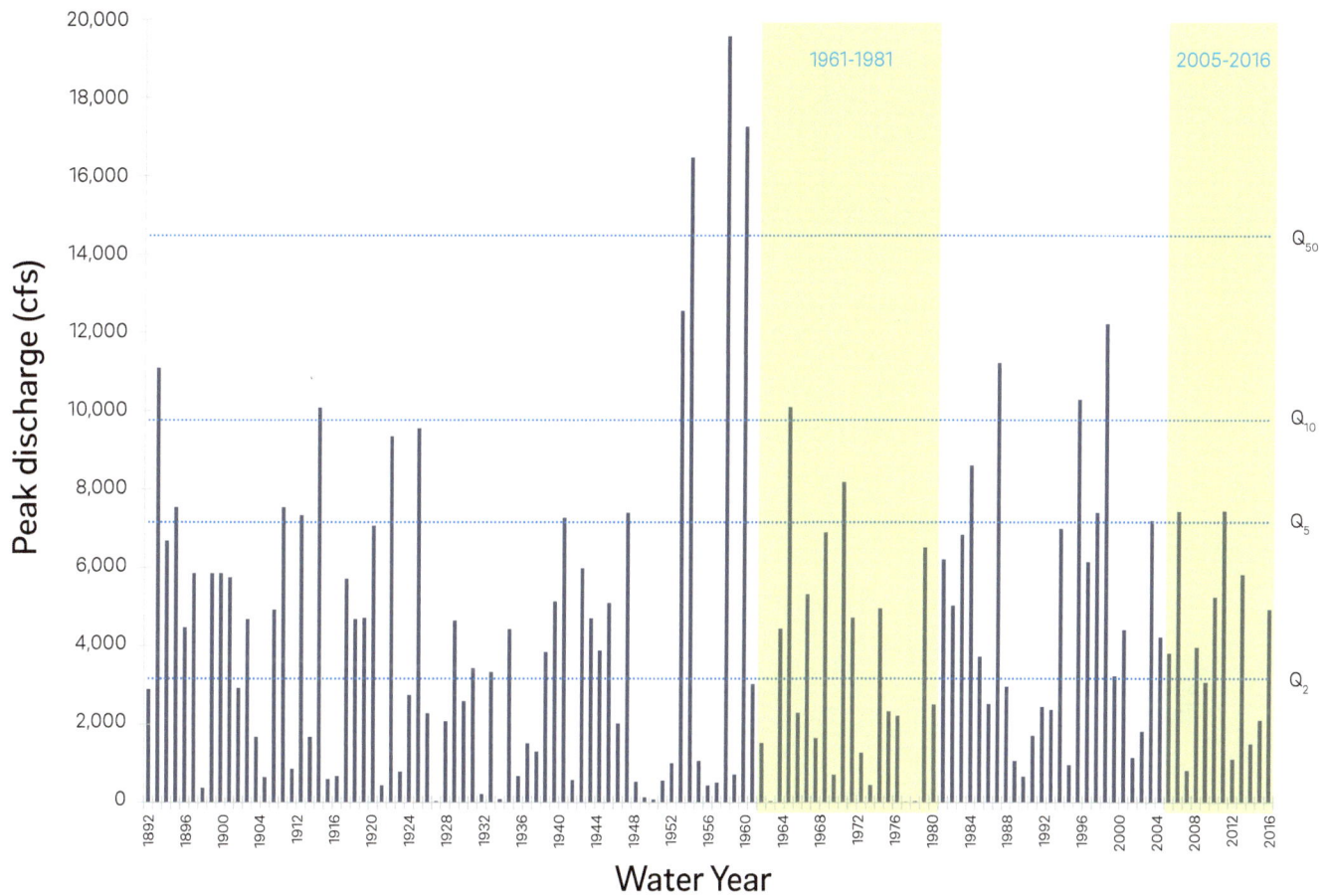

Figure 27. Annual peak discharges from Coyote Creek (1961-1981, 2005-2016) highlighted in green are directly from the USGS gaging station (11169800). All other dates use a regression from Alameda Creek at Niles gaging station (11179000).

TREE CORE AT PACHECO CREEK

45

Pacheco Creek Three medium-sized trees were cored along the inner floodplain of Pacheco Creek (Figure 28). The results of the tree ring analysis are shown in Table 12.

In contrast to the trees at Upper Coyote Creek, all three of the trees cored at Pacheco Creek did not have heart rot and it was possible to analyze complete cores for age. The trees cored at Pacheco were all relatively small, with diameter at breast height values under 10 inches. They also had straight trunks, and no snags, boles, or other complexities (Figure 29). It is possible that these trees may be hybrids, or may have been planted under a CalTrans mitigation program. The trees were, at a minimum, between 13 and 20 years of age. We did not hit pith at the center of the tree, and thus are not certain that we captured the true age of the trees. However, the ages of the trees cluster around the late 1990s and early 2000s. Large storms in that time period included the floods of 1995 and 1998. Unfortunately, no gaging data exists for this time period but large events occurred in these years throughout the region. According to our simple hydraulic analysis, the inner floodplain, where the trees are located, is an area that would be inundated above a 10-year flood event. Similar to Coyote Creek, the January 11, 2017 flood was a 25-year event on Pacheco Creek (11,400 cfs at the Pacheco Creek near Dunneville gage). It remains to be seen what, if any, regeneration is caused by this event.

While this pilot analysis was not conclusive, it lays the groundwork for understanding the role of flood disturbance in the successful recruitment and germination of sycamores in the recent past, suggesting conditions that may support conservation and restoration success in the future. Ultimately, such analyses have implications for the flow conditions necessary for the successful restoration of SAW habitat.

Table 12. Results of tree ring analysis on Pacheco Creek

Site	Tree ID	Length of core (cm)	Age (est)	Error (+/-)	% Error	Heart Rot	Minimum year est.
Pacheco	103	9.9	18	3	16.7	No	1998
Pacheco	93	14.3	13	5	38.5	No	2003
Pacheco	43	17.2	20	3	15.0	No	1996

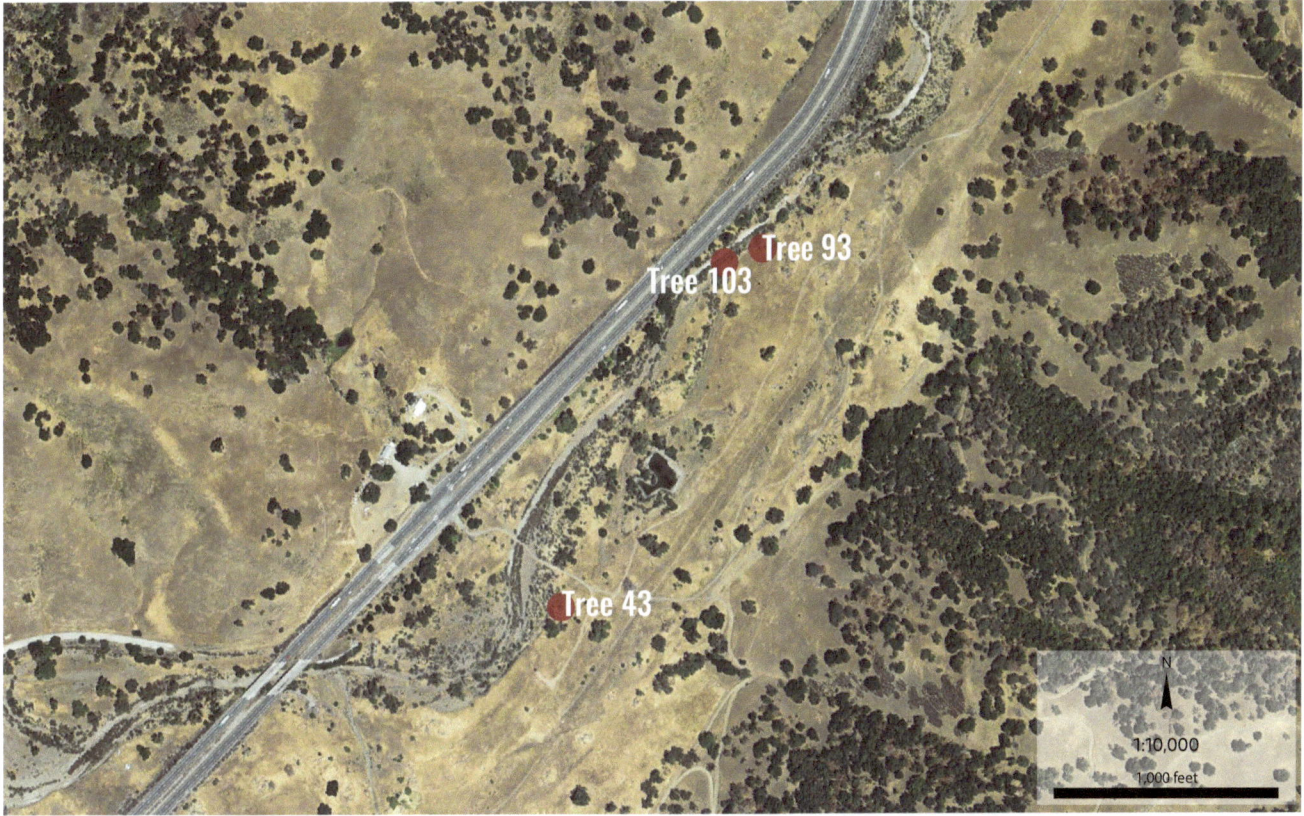

Figure 28. (Above) Locations of trees cored on Pacheco Creek. (NAIP 2014)

TREE 43 AT PACHECO CREEK

4 · DISCUSSION

SUMMARY OF RESULTS

In the following section, we revisit the hypotheses posited at the beginning of this study, addressing three major study questions, and reconcile them with the results derived from the field effort, anthracnose analysis, hydrologic assessment, tree coring, and literature review. In the "result" column, a green check mark indicates that the result of the study matched the hypothesis, a question mark indicates that the result was mixed or inconclusive, and a red X-mark indicates that the result did not match the hypothesis.

SYCAMORE SNAG WITH WOODPECKER GRANARY, UPPER COYOTE CREEK

HOW ARE SYCAMORES DISTRIBUTED AT EACH SITE?

	Expected Results in "Natural" System [UPPER COYOTE]	Result	Expected Results in "Managed" System [PACHECO]	Result
By geo-morphic zone	Younger trees are concentrated along the primary channel, older trees are concentrated in the older floodplain.	?	Younger trees are concentrated along the primary channel, older trees are concentrated in the older floodplain.	✓
By distance from channel	Younger trees are located closer to the channel, older trees are located further.	?	Younger trees are located closer to the channel, older trees are located further.	✓

DISTRIBUTION • In general, we expected to find a positive relationship between size class and distance to the primary channel. This is what we found at Pacheco - young trees were most concentrated close to the channel and in the inner floodplain and primary channel geomorphic zones, and larger, older trees were found at relative higher densities further from the channel. This is consistent with literature that suggests sycamores develop on newly-formed alluvial deposits near the channel (Stromberg 2001). At Upper Coyote Creek, we did not observe this pattern. This was perhaps due to historic migration of the channel over time. However, the distribution of large trees at Upper Coyote may reflect a historical path of an older channel along the northern edge of the floodplain.

COBBLES AND MULEFAT AT PACHECO CREEK

WHAT IS THE GENERAL HEALTH OF SYCAMORES?

	Expected Results in "Natural" System [UPPER COYOTE]	Result	Expected Results in "Managed" System [PACHECO]	Result
By health / vigor score	Trees are relatively *healthier*.	?	Trees are relatively *less healthy*.	?
By mortality	Trees suffer *less mortality*.	✓	Trees suffer *greater mortality*.	✓
By anthracnose symptoms	Trees have *lower incidence*.	X	Trees have *greater incidence*.	X

HEALTH • In general, we expected to find healthier trees in the more dynamic, more flood prone, less hydrologically managed, and more "natural" system at Upper Coyote Creek. More frequent and intense flood events would be expected to promote conditions more conducive to sycamore growth - availability of cobble sediment, removal of anthracnose-infected litter, and thinning of competitor species. This was not definitively supported by all the metrics examined.

By the health/vigor score, the trees did not differ substantially across size class, geomorphic zone or on average between sites. Healthier trees were generally located closer to and concentrated near the active channel at both sites.

By the mortality score, trees suffered minimal mortality at Upper Coyote Creek – survivorship was around 99%. At Pacheco Creek, mortality was more pronounced, with close to 50% of large trees and around 15% of medium trees dead. The tributary and outer floodplain suffered the highest proportional densities of mortality. Further, dead trees were located further from the channel on average at both sites. The fact that mortality was much higher at the managed site is congruent with initial predictions. A more natural, dynamic site such as Upper Coyote Creek would be expected to experience higher survivorship. Furthermore, many of the dead trees at Pacheco Creek may be associated with Harper Creek, a tributary at the western edge of the site, but CDFW observations of general die-back of sycamores throughout the site remains unexplained. Groundwater pumping and stream diversions may be the cause of some of the mortality both along Harper Creek, and the main stem of Pacheco. Three stilling wells were found on the outer floodplain at Pacheco Creek; however we were unable to obtain the data associated with those wells. Groundwater levels are a critical element to the survival of sycamore trees, and further studies are recommended to assess the groundwater conditions at both sites.

By anthracnose incidence, Upper Coyote Creek suffered a larger proportion of trees with high-severity symptoms than Pacheco Creek. Symptom severity did not vary much across zone or tree size at either site. Further, results from tree coring data suggest many trees at Upper Coyote Creek trees suffer from heart rot. Anthracnose stress may have led to greater heart rot incidence at Upper Coyote Creek. It is possible that many of the sycamores at Pacheco Creek could be hybrids of

ANTHRACNOSE AFFECTED LEAVES

California sycamore and non-native London plane tree. London plane tree is thought to be less susceptible to sycamore anthracnose than California sycamore (Sinclair et al. 1987, Oswald 2002). London plane tree's resistance to sycamore anthracnose may result in less dead wood and trunk cavities compared with California sycamore (Johnson et al. 2016). Some studies have suggested that natural systems experience more intense, more frequent flood events to wash away anthracnose-infected litter (Kamman Hydrology 2009), which, by implication, would reduce local infestation intensity.

Sampling time frame may explain these results in part – these data points were collected in one season only. Perhaps across many seasons, anthracnose-infected litter would be generally washed away, reducing probability of local spread. Depending on litter distribution, Pacheco Creek may experience greater litter removal from less intense but more frequent events from the perennial flow, particularly if Pacheco Creek's channel is more incised, which is true in some locations. Upper Coyote Creek also does not appear to be experiencing much mortality, so perhaps there is simply high variability in local incidence – literature suggests that anthracnose as a stressor is rarely lethal in isolation (Stuart 2001, Crump 2009). However, these observations, combined with noted heart rot, may forecast an impending threat for many of the sycamores at Upper Coyote Creek. Differences in microclimate may also contribute to observed differentials in disease vectors. Many other factors affect tree health; further research will be required to understand the primary drivers of sycamore health and abundance.

HOW ARE SYCAMORES REGENERATING?

	Expected Results in "Natural" System [UPPER COYOTE]	Result	Expected Results in "Managed" System [PACHECO]	Result
By new growth	Significant *new generation.*	X	Relatively *less regeneration.*	X
By root sprouts	Trees have *higher root sprout production.*	?	Trees have *lower root sprout production.*	?
By seed production	Trees have *more seed production.*	?	Trees have *less seed production.*	?

DEAD SYCAMORE TREE AT PACHECO CREEK

REGENERATION • Regeneration at Pacheco Creek followed a predictable pattern of decreased newer growth with greater distance from the channel. Though we expected natural systems to support more regeneration at Upper Coyote Creek, this result was not supported, as Upper Coyote Creek experienced very little recruitment. This could be due to multiple factors. One potential explanation for this observed trend is differences in grazing management. There was cattle exclusion fencing at both Pacheco Creek and Upper Coyote Creek. While there were no signs of active grazing observed at Pacheco Creek, observations of cattle tracks and dung at Upper Coyote Creek indicate some degree of trespass grazing. Given the absence of more palatable competitors such as cottonwood (Shanfield 1984) and the new production (but not apparent persistence) of seedlings and suckers, cattle grazing may be limiting regeneration at Upper Coyote Creek. Another explanation for this differential in recruitment could be explained by anthracnose incidence. As discussed in the Health section, anthracnose incidence is more prevalent and severe at Upper Coyote. Anthracnose affects new leaves and has been hypothesized to affect seed production (Shanfield 1984), which suggests that high site anthracnose incidence may be a limiting factor for regeneration. One final potential explanation is the length of the study period; perhaps under longer time scales, we would observe more flood events and regeneration at Upper Coyote Creek. Conditions for establishment of sycamores are highly specific – flood events must deposit alluvial sediment, sycamores must produce and lay a successful seed set, with timing such that a high initial water table that allows for seedling establishment then has sufficient draw-down to allow root aeration (Keeler-Wolf 1996). Perhaps regeneration is currently infeasible due to the water table levels, but could resume with a large flood event (the last one at this site was 1998). This data "snapshot" likely provides an incomplete picture of full, dynamic regeneration patterns. Further study could include the seed production of sycamores upstream in the canyon above the Upper Coyote Creek site, which could provide seed sources in a large flood event.

Data from seed production and root sprouting provided mixed results. Overall, seed production and root sprouting did not seem to differ dramatically between sites. Seeds were present almost exclusively on medium or large trees at both sites. In general, the observed reproductive strategy was clonal regeneration, which is well supported by the literature (which confirms generation from seed is rare (Shanfield 1984, Bock and Bock 1989, Finn 1991, Stromberg 2002). There appeared to be no obvious pattern of substrate with new growth, though we expect new growth from seed to be on cobbles and near the channel. However, our metrics for substrate were coarse and could be more refined (by examining subsurface sediments) in future studies to better capture this association. There also was no obvious pattern of species associations with new growth.

5 • CONCLUSION, RECOMMENDATIONS, AND FUTURE STUDY

In conclusion, we describe several site characteristics that may be most favorable to the health and regeneration of SAW, suggest specific locations for land acquisition or SAW restoration, suggest modifications to existing management practices to benefit SAW, and identify opportunities to try new ways of re-establishing sycamores. Embedded in this is a discussion of many unanswered questions about the reproduction and health of sycamores that warrant further study at these sites and others.

SYCAMORE AT UPPER COYOTE CREEK

SITE CHARACTERISTICS AND OBSERVATIONS THAT MAY BE FAVORABLE TO HEALTH AND REGENERATION OF SAW

Sycamores appear to do well when they are near an active channel with intermittent flow. Pacheco Creek follows hypothesized patterns of sycamore distribution and regeneration, with smaller and younger trees located nearest to the active channel. Patterns of distribution at Upper Coyote Creek are harder to explain given the small sample size for younger size classes and a channel that has migrated over time. In general, healthier and living trees were located closer to the channel than dead or unhealthy trees, at both sites. However, it is possible that "unhealthy" sycamore trees (with cavities, heart rot, snags) may be more beneficial for wildlife than "healthy" trees, such as those we see a Pacheco Creek.

Natural hydrographs alone do not guarantee healthy regeneration, but flooding is necessary. While the hydrology of Upper Coyote Creek is less modified, Pacheco Creek is a much more varied site in terms of sycamore patterns: Pacheco Creek has experienced significant mortality and regeneration, while Upper Coyote Creek experiences very little (though trees at Upper Coyote contribute a more dispersed and more dense seed pool). This is also visible in the size class distribution – Upper Coyote Creek is mostly composed of long-lived, larger trees, while Pacheco Creek has a more even size/age distribution. However, such even-aged stands may result when large (>100 year) flood events remove old (possibly hollowed out and weakened) trees and a new, young, even-aged stand is generated. Such large flood events have not been seen in the past 65 years, and occur very infrequently. Pacheco Creek's flooding cycle may have been interrupted by the dam, but encroachment of dense woody vegetation (by mulefat) along the main channel may allow for a "nursery" like condition which protects sycamore seedlings from grazing and/or provides a favorable seed germination microclimate. We would expect reproduction to be more favorable where there are natural flows, fresh deposition, and organization of coarse sediment as we see at Upper Coyote Creek. However we observe more regeneration at Pacheco Creek, where there is an armored bed, a modified hydrograph, and more anthropogenic disturbance.

Anthracnose may affect California sycamore regeneration. At Upper Coyote Creek, anthracnose infestation severity was high and regeneration was low. In contrast, at Pacheco Creek, anthracnose infestation severity was low and regeneration was high. Site history at Pacheco is not fully known, and there is some question about whether the seedlings/suckers and saplings at Pacheco Creek were planted. The average health and vigor of sycamores was medium-high at both study sites. Thus, sycamore anthracnose may have little effect on the apparent health and vigor of California sycamores; however, anthracnose may reduce the overall reproductive fitness of California sycamore by limiting seed and root sprout production.

It is necessary to understand the effects of hybridization between California sycamore and London plane tree on California sycamore regeneration. Hybridization between California sycamore and London plane tree has been documented in both natural and horticultural settings and may result in the extinction of the native genotype and a loss of habitat values for wildlife (Byington 2016, Johnson et al. 2016). A genetics study is underway that will examine the nature of hybridization between California sycamore and London plane tree at Pacheco Creek and Upper Coyote Creek and inform future studies on sycamore hybridization and regeneration. Other studies are proposed to analyze the physical structure of California sycamore, London plane tree, and hybrids to determine whether there are potential differences in wildlife habitat values to guide the management and restoration of SAW.

Regeneration success may be impacted by grazing and browsing. While not directly evaluated during this study, grazing and browsing may be a key limiting factor for sycamore regeneration. An experiment in exclusion fencing could inform management techniques at both sites. It is possible that the flow requirements are met at Upper Coyote Creek, but the seedlings and suckers are limited by grazing and browsing pressures and are not surviving. Gaining a better understanding about the difference in management between the two sites, and doing exclusion-fence testing will be critical to understanding if seedlings are establishing at Upper Coyote Creek, but not surviving, or if the health and age of the trees has an impact on their ability to reproduce.

SPECIFIC LOCATIONS ARE RECOMMENDED FOR ACQUISITION, RESTORATION, OR MANAGEMENT WITHIN THE VHP PRIORITY PRESERVE AREAS

Based on the local knowledge of HTH staff, and previous mapping efforts in the County, there are several sites preliminarily recommended for enhancement of SAW. To determine whether or not these sites are suitable for SAW enhancement, further site-specific investigations should examine the following parameters: location relative to the VHP Priority Preserve Areas; ownership; hydrology (managed versus natural); range of geomorphic zones; livestock grazing; and potential for acquisition, restoration, or management. We recommend the following locations be considered (Figure 30):

- Calero Creek (below dam)
- Upper Coyote Creek (from Anderson Dam to Hellyer Park)
- Guadalupe Creek
- Hellyer County Park
- Coe Park- Hunting Hollow parking lot to approximately 0.5 miles upstream
- Llagas Creek (outside SCVWD project/mitigation area) from Morgan Hill to San Martin
- Pacheco Creek (upstream/downstream of Caltrans site)
- Springbrook mitigation site on Norwood Creek
- Upper Penitencia Creek
- Uvas Creek

Compilation of Sycamore Alluvial Woodland Sites in Santa Clara County

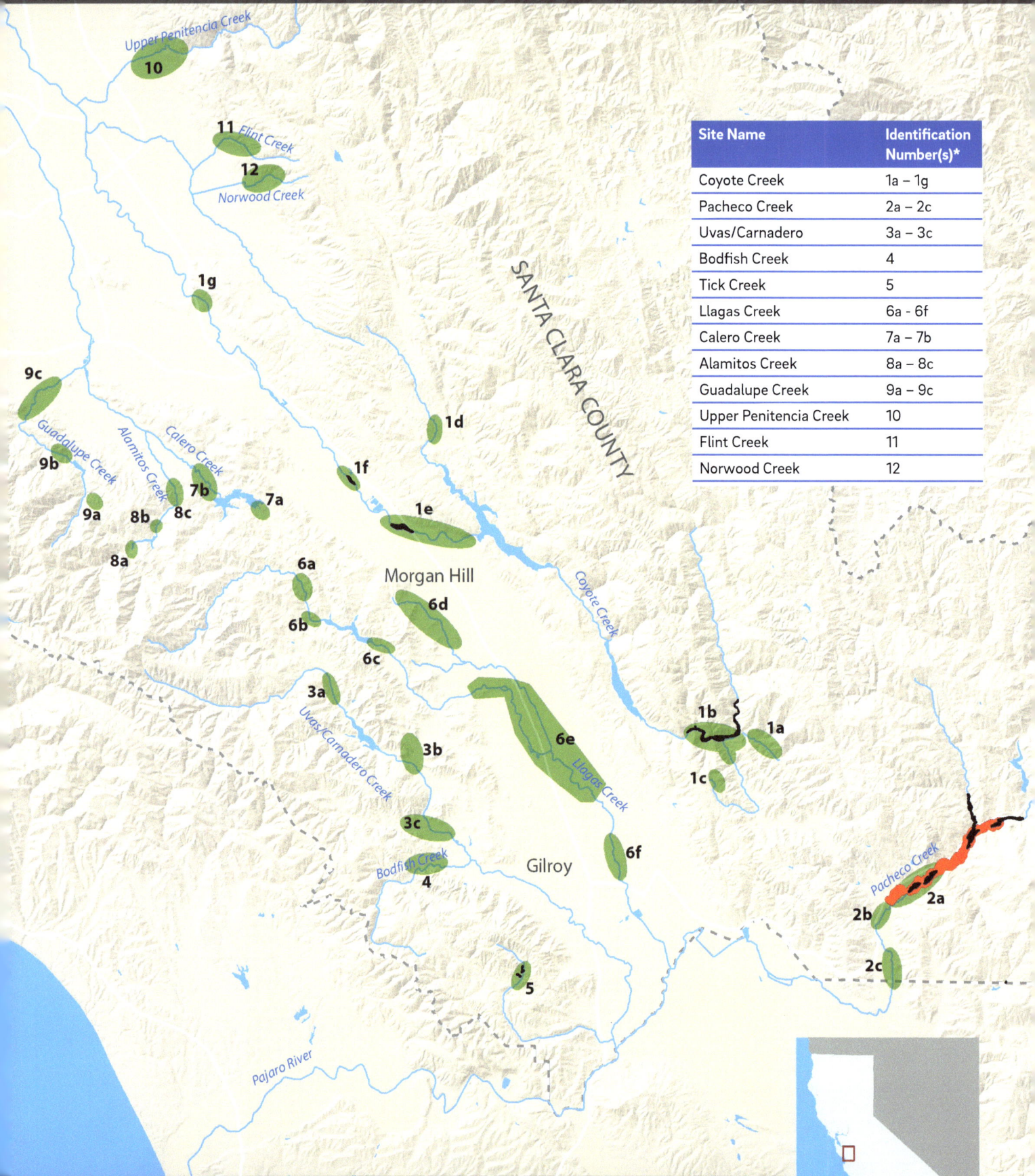

San Jose

Upper Penitencia Creek

10

11 Flint Creek

12

Norwood Creek

SANTA CLARA COUNTY

1g

9c

1d

Guadalupe Creek

9b

Calero Creek

1f

Alamitos Creek

7b

7a

1e

9a

8b

8c

8a

6a

Morgan Hill

6b

6d

Coyote Creek

6c

3a

Uvas/Carnadero Creek

3b

6e

1b

1a

Llagas Creek

1c

3c

Bodfish Creek

6f

2a

4

Gilroy

2b

5

2c

Pajaro River

Site Name	Identification Number(s)*
Coyote Creek	1a – 1g
Pacheco Creek	2a – 2c
Uvas/Carnadero	3a – 3c
Bodfish Creek	4
Tick Creek	5
Llagas Creek	6a - 6f
Calero Creek	7a – 7b
Alamitos Creek	8a – 8c
Guadalupe Creek	9a – 9c
Upper Penitencia Creek	10
Flint Creek	11
Norwood Creek	12

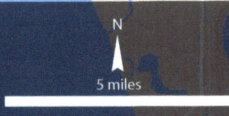

Pacheco Creek

N

5 miles

Valley Habitat Plan (Santa Clara County 2015)

California Department of Fish and Game (1996)

Sycamore stands to consider for enhancement (observed HTH 2016)

HOW CAN CURRENT MANAGEMENT PRACTICES BE MODIFIED TO SUPPORT SAW?

From this study, several recommendations to consider for site-specific practices supporting SAW at Upper Coyote and Pacheco Creeks are discussed below.

Use pulsed flows from reservoir at Pacheco Creek. By further studying the flows required to create the geomorphic disturbance that spurs sycamore regeneration, restoration managers may be able to provide more beneficial hydrology to sites such as Pacheco Creek, which is just downstream from an under-used reservoir upstream. Initial results from the tree coring and hydrologic analysis suggest that a minimum 10-25 year flood event may be necessary to activate regeneration of sycamores at Pacheco Creek. Further study should focus on testing this early finding and monitoring the sites carefully after such large flows do occur.

Modify grazing patterns at Upper Coyote Creek to limit herbivory on young sycamore trees. Restoration managers should consider modifying grazing patterns at both sites, but particularly at Upper Coyote Creek where seedlings/saplings either do not establish, or do not survive. Creating experiment plots with exclusion fencing may be a way to test whether survivorship is the issue. Pacheco Creek also has had a varied grazing history, but seedlings seem to be surviving with higher success rates (perhaps because they are protected by woody plant "nurseries").

Control woody invasive species that grow in geomorphic zones where sycamore are most likely to regenerate. As observed at Pacheco Creek, sycamore seedlings and saplings appeared to thrive along the primary channel and inner channel corridor, often within stands of mulefat. Other woody species such as tamarisk or tree of heaven may compete with sycamores for light and space, and could be controlled to preferentially support sycamore survival within primary channels and inner channel corridors.

RECOMMENDATIONS FOR ACTIVE RESTORATION METHODOLOGIES OF SAW

Several concepts for actively restoring and managing SAW stands are discussed below:

Plant California sycamore in inner channel geomorphic zones along tributaries. California sycamore prefers to grow/establish along intermittent stream channels on low terraces of the inner channel corridor and floodplain. These zones, where water availability is high relative to other geomorphic zones, may be appropriate places to install container stock and/or cuttings. California sycamore propagule sources should be of pure genetic stock to ensure that sycamore hybrids are not planted. The restoration plantings should be installed between October and December when rainfall has saturated the soils. Irrigate the plantings during hot summer months to help them become established, especially in years with below average rainfall. Control weeds throughout the planting area to reduce competition with invasive plants. HTH and SFEI are developing a pilot restoration project to test different techniques that could be used to establish California sycamores, as described below under "Further investigations."

Use exclusionary fencing for grazing. Experimentally exclude livestock from portions of SAW habitats (e.g. within the inner channel corridors) to test the effects of grazing on the natural recruitment of California

Figure 30. Compilation of sycamore sites in Santa Clara County, including mapping by CDFW, the Santa Clara Valley VHP, and observations by HTH staff.

sycamore. Exclude livestock from portions of the restoration site to test the effects of grazing on California sycamore establishment. Examine how differences in soil compaction, herbaceous biomass, and damage to leaves and stems affect the natural recruitment and establishment of California sycamore.

Improve methods to propagate California sycamore planting stock for restoration projects. Propagation of California sycamore for restoration projects is typically done using seed. However, the plants produced using this technique often are hybrids between California sycamore and London plane tree. To ensure that pure California sycamore plants are available for restoration projects, they must be propagated vegetatively from pure stock. Vegetative propagation of California sycamore is extremely difficult. Only approximately 10% of the cuttings taken ultimately result in a plant that can be used for restoration purposes. Vegetative propagation studies should seek to improve on this poor performance to advance the science of vegetative propagation of California sycamore and to improve the cost-effectiveness of vegetative propagation of California sycamore. Such a study, funded by the Santa Clara Valley Water District (SCVWD), is underway (see below, under Next Steps).

KEY UNCERTAINTIES

Many uncertainties remain surrounding the distribution, health, and regeneration patterns of sycamores which could be studied in the future.

Relationship of flooding to sycamore regeneration. More work should be done to understand the magnitude and timing of floods and their relationship to sycamore regeneration. This will be essential for influencing management actions like reservoir pulses. This could be accomplished by investigating the ages and locations of more sycamores using dendrochronology methods, observing sites before and after flood events, and tracking the location and timing of new seedlings and saplings.

Groundwater conditions. A major uncertainty not investigated in this study relates to the groundwater conditions necessary to sustain SAW stands. The relationship between groundwater conditions and sycamore regeneration and survival at the two sites, as well as others in Santa Clara Valley is not well known. Future studies should consider installing piezometers to track groundwater dynamics over time, as well as incorporating data that has already been collected at Pacheco Creek.

Plant pathogens. Uncertainties exist concerning the effects of the plant pathogen *Phytophthora megasperma* on sycamore regeneration. *Phytophthora* is a fungus-like watermold found in plant nurseries, landscapes, agricultural fields, and native habitats that can damage and kill host plants. It infects leaves, stems, and roots, causing leaf blight, cankers, and branch die-back. *Phytophthora* can invade the water conducting wood (xylem) of trees beneath the inner bark, and cause water stress symptoms in all or part of the canopy. In California, introduced *Phytophthora* has affected a number of plant communities in a variety of soils and climates (Swiecki and Bernhardt 2014). California sycamore is a host for *Phytophthora*. Very little is known about how this species affects California sycamore regeneration, or whether California sycamore is a host for other species of *Phytophthora*.

Hybridization of sycamores. Uncertainties also exist concerning the effects of California sycamore hybridization with London plane tree on regeneration of California sycamore. London plane tree is thought to be less susceptible to sycamore anthracnose than California sycamore (Sinclair et al. 1987, Oswald 2002). Thus, it is reasonable to assume that hybrid trees would be less susceptible to anthracnose than

California sycamore. Anthracnose infestations can cause tree limbs to die and trunk cavities to form (Sinclair et al. 1987). Such dead wood and cavities provide important foraging substrate and nesting sites for birds; however, severe damage by anthracnose may weaken trees and contribute to a lack of regeneration. Further investigation is needed to determine how hybridization between California sycamore and London plane tree may affect California sycamore regeneration and the habitat values of California sycamores and sycamore hybrids in riparian habitat. Hybridization between the California sycamore and the London plane tree has become a major challenge to restoration goals and potentially threatens the continued genetic distinctiveness of this native species (Johnson et al. 2016; see also Next Steps, below, for genetics study currently underway). For these reasons, local restoration nurseries have attempted to find a viable approach for reproducing non-hybrid California sycamore plant material. Unfortunately, seed collected from large, seemingly pure California sycamore has been observed to produce seedlings that by leaf morphology are thought to be hybrids, thus frustrating efforts to use seed to produce native stock. Vegetative propagation of California sycamore has proven to be very challenging, with most attempts yielding few if any viable seedlings that can be used for restoration, but it remains the only means of reliably producing native stock. Further investigations will seek to advance the science of vegetative propagation of California sycamore and improve the cost-effectiveness of vegetation propagation of California sycamore.

NEXT STEPS

Because of increasing interest in restoring this rare and important habitat type, several efforts are already underway.

HTH is studying the hybridization between California sycamore and London plane tree. Funded by the SCVWD, HTH is in the midst of a genetics study designed to develop a better understanding of the nature of hybridization between California sycamore and London plane tree in southern Santa Clara County and to locate California sycamore individuals that can be used as propagule sources for propagation studies and restoration projects. The genetics study involves analyzing leaf samples from a combination of California sycamore, London plane tree, and potential hybrids to determine the ancestry fraction (percent native California sycamore) in each tree sampled and to identify genetically pure California sycamore "mother" trees. The results of SFEI's tree coring and the genetic analysis will help determine approximately when hybridization began to occur in southern Santa Clara County. If a point in time can be identified before which hybridization did not occur, then we will identify the minimum tree size (diameter at breast height) that can be used as a "rule of thumb" to select pure California sycamore trees as source materials for propagation. Such a short cut would be a significant advantage for future sycamore restoration projects.

HTH is leading a propagation study to advance the science of vegetative propagation of California sycamore and improve the cost-effectiveness of vegetative propagation of California sycamore. Funded by SCVWD project funds and a SCVWD grant, the propagation experiment will test a number of treatments, including: cutting material (basal sprouts and stems), cutting preparation (simple cut and heal cut), willow water presoak, rooting media, and cutting season (spring, fall). Response variables for assessing California sycamore cutting success will include: survival, growth, and vigor. A report will be prepared that will provide clear direction to the reader regarding the most promising techniques that can be used. It also will include clear recommendations concerning further studies that could be performed based on knowledge gained from this study.

A study will be designed for a pilot restoration project to test different techniques that could be used to establish California sycamore. Funded by a SCVWD grant, the experimental design will involve biological, geomorphic, and cartographic work that will build on the knowledge gained from this habitat mapping and regeneration study. Using existing data, HTH and SFEI will continue their investigation of the relationship between flows, geomorphic features, and SAW stand age and establishment at Upper Coyote Creek and Pacheco Creek. The pilot restoration project design will potentially include elements such as planting sycamore trees in different geomorphic zones, planting at different distances from the main creek channel, and planting with different soil and vegetation types. We also will experiment with planting California sycamore cuttings directly into the ground and will compare the results with those obtained from planting container stock. The study will measure signs of establishment and growth, such as survival, canopy spread, diameter, height, health, and vigor.

References

Barnes, H.H. Jr. 1967. Roughness Characteristics of Natural Channels. Geological Survey Water-Supply Paper 1849. United States Government Printing Office, Washington.

Beagle, J. S. Baumgarten, R. M. Grossinger, R. A. Askevold, B. Stanford. 2014. Landscape scale management strategies for Arroyo Mocho and Arroyo Las Positas: process-based approaches for dynamic, multi-benefit urban channels. SFEI Publication #714, San Francisco Estuary Institute, Richmond, California.

Belli, J. P. 2015. Movements, Habitat Use, and Demography of Western Pond Turtles in an Intermittent Central California Stream. Master's Thesis. San Jose State University.

Bock, C. E., and J.H. Bock. 1981. "Importance of sycamores to riparian birds in southeastern Arizona." Journal of Field Ornithology 55, no. 1 (1984): 97-103.

Bock, J. H. and C. E. Bock. 1989. "Factors limiting sexual reproduction in Platanus wrightii in southeastern Arizona. "Aliso 12:295–301.

Byington, C. 2016. Unraveling the Mystery of the Western Sycamores that Weren't. March 11. The Nature Conservancy. Arlington, Virginia. <http://www.nature.org/>.

[Caltrans] California Department of Transportation. 1996. Army Permit No. 16477S92A SCL 152 Pacheco Creek Road Mitigation Monitoring Report for 1996. December 13. Oakland California. Prepared for U.S. Army Corps of Engineers. San Francisco, California.

Casagrande, J. M. 2010. Distribution, abundance, growth, and habitat use of steelhead in Uvas Creek, CA. Master's Thesis, San Jose State University.

Crump, A. 2009. Anthracnose. Univ. Calif. A. N. R. Pest Notes Publication 7420, Oakland, CA.

Environmental Laboratory. 1987. U.S. Corps of Engineers Wetlands Delineation Manual. Department of the Army.

Finn, M. S. 1991. Ecological characteristics of California sycamore Platanus racemosa. Thesis. California State University, Los Angeles.

Grossinger, R., E. Beller, M. Salomon, A. Whipple, R. Askevold, C. J. Striplen, E. Brewster, and R. A. Leidy. 2008. "South Santa Clara Valley Historical Ecology Study, including Soap Lake, the Upper Pajaro River, and Llagas, Uvas-Carnadero, and Pacheco Creeks." *Final report to the Santa Clara Valley Water District and Nature Conservancy. San Francisco Estuary Institute. Oakland, CA.*

Gillies, Eric L. 1998. "Effects of regulated stream flows on the Sycamore Alluvial Woodland riparian community."

Holland, R. 1986. Preliminary Descriptions of the Terrestrial Natural Communities of California. California Department of Fish and Game, Natural Heritage Division. Sacramento.

Johnson, M. G., K. Lang, P. Manos, G. H. Golet, and K. A. Schierenbeck. 2016. Evidence for genetic erosion of a California native tree, *Platanus racemose*, via recent, ongoing introgressive hybridization with an introduced ornamental species. Conservation Genetics 17:593–602.

Kamman Hydrology 2009. Phase 2 Technical Report Sycamore Grove Recovery Program. Sycamore Grove Park, Livermore, California. Prepared for Livermore Area Recreation and Park District and Zone 7 Water Agency.

King, J. 2004. Sycamore Grove Park dendrochronological investigation. Lone Pine Research, Bozeman, Montana.

Keeler-Wolf, T., K. Lewis, and C. Roye. 1996. The Definition and Location of Central California Sycamore Alluvial Woodland. Prepared by Natural Heritage Division, Bay-Delta and Special Water Projects Division, California Department of Fish and Game. May 1996. 111 pp. + appendices.

Lichvar, R. W., D. L. Banks, W. N. Kirchner, and N. C. Melvin. 2016. The National Wetland Plant List: 2016 wetland ratings. Phytoneuron 2016-30:1–17.

Oswald, V. H. 2002. Selected Plants of Northern California and Adjacent Nevada. Studies from the Herbarium. California State University, Chico.

San Francisco Estuary Institute (SFEI). 2015. Riparian Zone Estimation Tool Hydrologic Connectivity Module: Documentation and Validation of Selected Methodology. San Francisco Estuary Institute: 745.

Sawyer, J.O., T. Keeler-Wolf, and J. Evens. 2009. *Manual of California vegetation*. California Native Plant Society Press.

Shanfield, A. 1984. Alder, cottonwood, and sycamore distribution and regeneration along the Nacimiento River, California. Pages 196–201 *in* R. Warner and K. M. Hendrix, editors. California Riparian Systems. University of California Press. Berkeley, California.

Sinclair, W. A., H. H. Lyon, and W. T. Johnson. 1987. Diseases of Trees and Shrubs. Comstock Publishing Associates, a division of Cornell University Press. Ithaca, New York.

Stromberg, J. C. 2001. "Influence of Stream Flow Regime and Temperature on Growth Rate of the Riparian Tree, Platanus wrightii, in Arizona." Freshwater Biology 46:227–239.

Stuart, J. D. and J. O. Sawyer. 2001. Trees and Shrubs of California. University of California Press, Berkeley/Los Angeles/London.

Swiecki, T. and E. Bernhardt. 2014. Phytophthora species move from native plant nurseries into restoration plantings [presentation]. <http://caforestpestcouncil.org/wp-content/uploads/2014/12/Swiecki.pdf>.

Sycamore Associates, 2004. Sycamore Grove recovery program: phase 1 technical report, preliminary findings, Sycamore Grove Park, Livermore, Alameda County, California, final. Prepared for the Livermore Area Recreation and Park District, Livermore, CA. and Zone 7 Water Agency, Pleasanton, California, prepared by Sycamore Associates, Walnut Creek, CA.

Whitlock, D. L. 2003. The Hybridization of California Sycamore (*Platanus racemosa*) and the London Plane Tree (*Platanus x acerifolia*) in California's Riparian Woodland. Thesis. California State University, Chico.

Wolman, M. G. 1954. A method of sampling coarse river bed material. EOS, Transactions American Geophysical Union, 35(6), 951-956.

www.ingramcontent.com/pod-product-compliance
Lightning Source LLC
Chambersburg PA
CBHW041933220326
41598CB00058BA/835